Nuclear Engineering Handbook

Nuclear Engineering Handbook

Lindsay Garfield

Larsen & Keller
www.larsen-keller.com

Nuclear Engineering Handbook
Lindsay Garfield
ISBN: 978-1-64172-658-0 (Hardback)

 Larsen & Keller

Published by Larsen and Keller Education,
5 Penn Plaza,
19th Floor,
New York, NY 10001, USA

Cataloging-in-Publication Data

Nuclear engineering handbook / Lindsay Garfield
       p. cm.
Includes bibliographical references and index.
ISBN 978-1-64172-658-0
1. Nuclear engineering. 2. Engineering. 3. Nuclear energy. 4. Nuclear physics. I. Garfield, Lindsay.
TK9145 .N83 2022
621.48--dc23

For more information regarding Larsen and Keller Education and its products, please visit the publisher's website www.larsen-keller.com

Table of Contents

Permissions

Index

Preface

The branch of engineering which deals with the application of different sub-atomic processes based on the principles of nuclear physics is known as nuclear engineering. The two primary sub-atomic processes, which are dealt within this field are fission and fusion. Fission refers to the breaking down of atomic nuclei while fusion refers to the combining of atomic nuclei. The field of nuclear fission deals with designing, maintaining and interacting with components and systems like nuclear weapons and nuclear reactors. It also involves the study of medical and other applications of radiation, especially ionizing radiation, heat/thermodynamics transport, nuclear fuel, nuclear safety, and the problems of nuclear proliferation. The topics included in this book on nuclear engineering are of utmost significance and bound to provide incredible insights to readers. The various studies that are constantly contributing towards advancing technologies and evolution of this field are examined in detail. This book will serve as a reference to a broad spectrum of readers.

A short introduction to every chapter is written below to provide an overview of the content of the book:

Chapter 1 - The branch of engineering that deals with the breaking down and combining nuclei of atoms is referred to as nuclear engineering. Some of its concepts are nuclear physics, nuclear chemistry, nuclear reaction, radioactivity, etc. This is an introductory chapter which will briefly introduce all these significant concepts of nuclear engineering; **Chapter 2 -** A type of nuclear reaction in which energy is produced by combining two nuclei is called nuclear fusion whereas a nuclear reaction in which energy is produced by splitting of a single nucleus is called nuclear fission. This chapter has been carefully written to provide an easy understanding of nuclear fusion and fission; **Chapter 3 -** Substances required to carry out a nuclear reaction are termed as nuclear materials. Some of the materials that are used in nuclear reaction are uranium, plutonium, thorium, and their isotopes. This chapter closely examines these nuclear materials to provide an extensive understanding of this subject; **Chapter 4 -** A device which is used for initiation and maintenance of a self-sustained nuclear chain reaction is referred to as a nuclear reactor. It is used in generating electricity and nuclear marine propulsion. This chapter delves into the various types of nuclear reactor and their designs to provide an extensive understanding of this subject; **Chapter 5 -** Nuclear fuel is one of the important components of a nuclear reactor. Some of its aspects include fabrication of nuclear fuel, energy density calculations of nuclear fuel, nuclear fuel cycle, reprocessing of nuclear fuel, etc. This chapter closely examines these aspects of nuclear fuel manufacturing and reprocessing, to provide an extensive understanding of the subject; **Chapter 6 -** Radioactive waste refers to nuclear material which is left after a nuclear reaction. Radioactive waste management involves the treatment, storage, and disposal of liquid, airborne, and solid effluents from the nuclear industry's operations. The topics elaborated in this chapter will help in gaining a better perspective about radioactive waste management; **Chapter 7 -** Nuclear safety deals with rules of conduct for the safety of the workers and nuclear security refers to the prevention and detection of theft, inappropriate access, damage, etc. to the nuclear facility. Some of the fields that fall under its domain are nuclear accidents, nuclear safety, nuclear criticality safety, nuclear reactor safety, nuclear power plant safety systems, etc.

This chapter has been carefully written to provide an easy understanding of these fields which associates with nuclear safety and security.

I extend my sincere thanks to the publisher for considering me worthy of this task. Finally, I thank my family for being a source of support and help.

Lindsay Garfield

Nuclear Engineering: An Introduction

The branch of engineering that deals with the breaking down and combining nuclei of atoms is referred to as nuclear engineering. Some of its concepts are nuclear physics, nuclear chemistry, nuclear reaction, radioactivity, etc. This is an introductory chapter which will briefly introduce all these significant concepts of nuclear engineering.

Nuclear Power

Nuclear power is the electricity generated by power plants that derive their heat from fission in a nuclear reactor. Except for the reactor, which plays the role of a boiler in a fossil-fuel power plant, a nuclear power plant is similar to a large coal-fired power plant, with pumps, valves, steam generators, turbines, electric generators, condensers, and associated equipment.

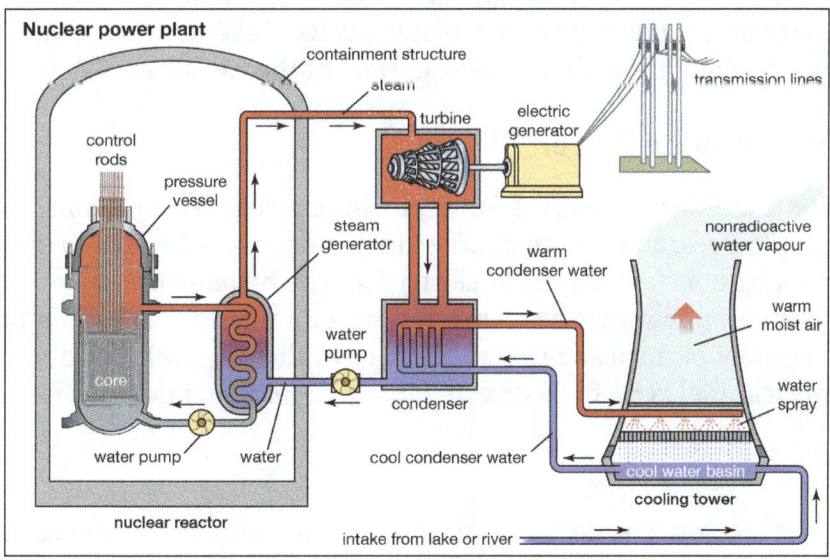

Nuclear power provides almost 15 percent of the world's electricity. The first nuclear power plants, which were small demonstration facilities, were built in the 1960s. These prototypes provided "proof-of-concept" and laid the groundwork for the development of the higher-power reactors that followed.

The nuclear power industry went through a period of remarkable growth until about 1990, when the portion of electricity generated by nuclear power reached a high of 17 percent. That percentage remained stable through the 1990s and began to decline slowly around the turn of the 21st century, primarily because of the fact that total electricity generation grew faster than electricity from nuclear power while other sources of energy (particularly coal and natural gas) were able to grow more quickly to meet the rising demand. This trend appears likely to continue well into the 21st century. The Energy Information Administration (EIA), a statistical arm of the U.S. Department of Energy, has projected that world electricity generation between 2005 and 2035 will roughly double (from more than 15,000 terawatt-hours to 35,000 terawatt-hours) and that generation from all energy sources except petroleum will continue to grow.

In 2012 more than 400 nuclear reactors were in operation in 30 countries around the world, and more than 60 were under construction. The United States has the largest nuclear power industry, with more than 100 reactors; it is followed by France, which has more than 50. Of the top 15 electricity-producing countries in the world, all but two, Italy and Australia, utilize nuclear power to generate some of their electricity. The overwhelming majority of nuclear reactor generating capacity is concentrated in North America, Europe, and Asia. The early period of the nuclear power industry was dominated by North America (the United States and Canada), but in the 1980s that lead was overtaken by Europe. The EIA projects that Asia will have the largest nuclear capacity by 2035, mainly because of an ambitious building program in China.

A typical nuclear power plant has a generating capacity of approximately one gigawatt (GW; one billion watts) of electricity. At this capacity, a power plant that operates about 90 percent of the time (the U.S. industry average) will generate about eight terawatt-hours of electricity per year. The predominant types of power reactors are pressurized water reactors (PWRs) and boiling water reactors (BWRs), both of which are categorized as light water reactors (LWRs) because they use ordinary (light) water as a moderator and coolant. LWRs make up more than 80 percent of the world's nuclear reactors, and more than three-quarters of the LWRs are PWRs.

Issues Affecting Nuclear Power

Countries may have a number of motives for deploying nuclear power plants, including a lack of indigenous energy resources, a desire for energy independence, and a goal to limit greenhouse gas emissions by using a carbon-free source of electricity. The benefits of applying nuclear power to these needs are substantial, but they are tempered by a number of issues that need to be considered, including the safety of nuclear reactors, their cost, the disposal of radioactive waste, and a potential for the nuclear fuel cycle to be diverted to the development of nuclear weapons.

Safety

The safety of nuclear reactors has become paramount since the Fukushima accident of 2011. The lessons learned from that disaster included the need to (1) adopt risk-informed regulation, (2) strengthen management systems so that decisions made in the event of a severe accident are based on safety and not cost or political repercussions, (3) periodically assess new information on risks posed by natural hazards such as earthquakes and associated tsunamis, and (4) take steps to mitigate the possible consequences of a station blackout.

The four reactors involved in the Fukushima accident were first-generation BWRs designed in the 1960s. Newer Generation III designs, on the other hand, incorporate improved safety systems and rely more on so-called passive safety designs (i.e., directing cooling water by gravity rather than moving it by pumps) in order to keep the plants safe in the event of a severe accident or station blackout. For instance, in the Westinghouse AP1000 design, residual heat would be removed from the reactor by water circulating under the influence of gravity from reservoirs located inside the reactor's containment structure. Active and passive safety systems are incorporated into the European Pressurized Water Reactor (EPR) as well.

Traditionally, enhanced safety systems have resulted in higher construction costs, but passive safety designs, by requiring the installation of far fewer pumps, valves, and associated piping, may actually yield a cost saving.

Economics

A convenient economic measure used in the power industry is known as the levelized cost of electricity, or LCOE, which is the cost of generating one kilowatt-hour (kWh) of electricity averaged over the lifetime of the power plant. The LCOE is also known as the "busbar cost," as it represents the cost of the electricity up to the power plant's busbar, a conducting apparatus that links the plant's generators and other components to the distribution and transmission equipment that delivers the electricity to the consumer.

The busbar cost of a power plant is determined by: 1) capital costs of construction, including finance costs, 2) fuel costs, 3) operation and maintenance (O&M) costs, and 4) decommissioning and waste-disposal costs. For nuclear power plants, busbar costs are dominated by capital costs, which can make up more than 70 percent of the LCOE. Fuel costs, on the other hand, are a relatively small factor in a nuclear plant's LCOE (less than 20 percent). As a result, the cost of electricity from a nuclear plant is very sensitive to construction costs and interest rates but relatively insensitive to the price of uranium. Indeed, the fuel costs for coal-fired plants tend to be substantially greater than those for nuclear plants. Even though fuel for a nuclear reactor has to be fabricated, the cost of nuclear fuel is substantially less than the cost of fossil fuel per kilowatt-hour of electricity generated. This fuel cost advantage is due to the enormous energy content of each unit of nuclear fuel compared to fossil fuel.

The O&M costs for nuclear plants tend to be higher than those for fossil-fuel plants because of the complexity of a nuclear plant and the regulatory issues that arise during the plant's operation. Costs for decommissioning and waste disposal are included in fees charged by electrical utilities. In the United States, nuclear-generated electricity was assessed a fee of $0.001 per kilowatt-hour to pay for a permanent repository of high-level nuclear waste. This seemingly modest fee yielded about $750 million per year for the Nuclear Waste Fund.

At the beginning of the 21st century, electricity from nuclear plants typically cost less than electricity from coal-fired plants, but this formula may not apply to the newer generation of nuclear power plants, given the sensitivity of busbar costs to construction costs and interest rates. Another major uncertainty is the possibility of carbon taxes or stricter regulations on carbon dioxide emissions. These measures would almost certainly raise the operating costs of coal plants and thus make nuclear power more competitive.

Radioactive-waste Disposal

Spent nuclear reactor fuel and the waste stream generated by fuel reprocessing contain radioactive materials and must be conditioned for permanent disposal. The amount of waste coming out of the nuclear fuel cycle is very small compared with the amount of waste generated by fossil fuel plants.

However, nuclear waste is highly radioactive (hence its designation as high-level waste, or HLW), which makes it very dangerous to the public and the environment. Extreme care must be taken to ensure that it is stored safely and securely, preferably deep underground in permanent geologic repositories.

Despite years of research into the science and technology of geologic disposal, no permanent disposal site is in use anywhere in the world. In the last decades of the 20th century, the United States made preparations for constructing a repository for commercial HLW beneath Yucca Mountain, Nevada, but by the turn of the 21st century, this facility had been delayed by legal challenges and political decisions. Pending construction of a long-term repository, U.S. utilities have been storing HLW in so-called dry casks aboveground. Some other countries using nuclear power, such as Finland, Sweden, and France, have made more progress and expect to have HLW repositories operational in the period 2020–25.

Proliferation

The claim has long been made that the development and expansion of commercial nuclear power led to nuclear weapons proliferation, because elements of the nuclear fuel cycle (including uranium enrichment and spent-fuel reprocessing) can also serve as pathways to weapons development. However, the history of nuclear weapons development does not support the notion of a necessary connection between weapons proliferation and commercial nuclear power.

The first pathway to proliferation, uranium enrichment, can lead to a nuclear weapon based on highly enriched uranium. It is considered relatively straightforward for a country to fabricate a weapon with highly enriched uranium, but the impediment historically has been the difficulty of the enrichment process. Since nuclear reactor fuel for LWRs is only slightly enriched (less than 5 percent of the fissile isotope uranium-235) and weapons need a minimum of 20 percent enriched uranium, commercial nuclear power is not a viable pathway to obtaining highly enriched uranium.

The second pathway to proliferation, reprocessing, results in the separation of plutonium from the highly radioactive spent fuel. The plutonium can then be used in a nuclear weapon. However, reprocessing is heavily guarded in those countries where it is conducted, making commercial reprocessing an unlikely pathway for proliferation. Also, it is considered more difficult to construct a weapon with plutonium versus highly enriched uranium.

More than 20 countries have developed nuclear power industries without building nuclear weapons. On the other hand, countries that have built and tested nuclear weapons have followed other paths than purchasing commercial nuclear reactors, reprocessing the spent fuel, and obtaining plutonium. Some have built facilities for the express purpose of enriching uranium; some have built plutonium production reactors; and some have surreptitiously diverted research reactors to the production of plutonium. All these pathways to nuclear proliferation have been more effective, less expensive, and easier to hide from prying eyes than the commercial nuclear power route. Nevertheless, nuclear proliferation remains a highly sensitive issue, and any country that wishes to launch a commercial nuclear power industry will necessarily draw the close attention of oversight bodies such as the International Atomic Energy Agency.

Nuclear Engineering

Nuclear engineering is the field engineering, the field of engineering that deals with the science and application of nuclear and radiation processes. These processes include the release, control, and utilization of nuclear energy and the production and use of radiation and radioactive materials for applications in research, industry, medicine, and national security. Nuclear engineering is based on fundamental principles of physics and mathematics that describe nuclear interactions and the transport of neutrons and gamma rays. These phenomena in turn are dependent on heat transfer, fluid flow, chemical reactions, and behaviour of materials when subjected to radiation. Nuclear engineering is therefore inherently a multifaceted discipline, relying on several branches of physics, and, like the aerospace industry, it relies to a large extent on modeling and simulation for the design and analysis of complex systems that are too large and expensive to be tested.

Installation of the dome of a containment structure at the
Taishan nuclear power plant.

Nuclear Engineering Functions

The functions carried out by nuclear engineers span a spectrum of activities, including basic research, applied research, development, fabrication and construction, operation, and product support and marketing. All of these functions apply to nuclear power and, to lesser degrees, to other branches of nuclear engineering, depending on the maturity of the field.

Nuclear engineers perform these functions for various categories of employers:

- Architectural engineering firms, for which they handle design, safety analysis, project coordination, construction supervision, quality assurance, quality control, and related matters.

- Reactor vendors and other manufacturing organizations, for which they pursue research, development, design, manufacture, and installation of various components of nuclear systems.

- Electric utility companies, for which they handle planning, construction supervision, reactor-safety analysis, in-core nuclear fuel management, power-reactor economic analysis, environmental-impact assessment, personnel training, plant management, operation-shift supervision, radiation protection, spent-fuel storage, and radioactive-waste management.

- Hospital and medical centres, where they conduct applied research and also develop and carry out diagnostic and therapeutic radiation procedures on patients.

- Regulatory agencies, for which they undertake licensing, rule making, safety research, risk analysis, on-site inspection, and research administration.

- Defense programs, for which they are employed in naval propulsion and nuclear weapons programs.

- Universities, where they teach prospective nuclear engineers and perform basic and applied nuclear engineering research.

- National laboratories and industrial research laboratories, where they carry out basic and applied research and development on a variety of nuclear-related topics.

Branches of Nuclear Engineering

Nuclear Power

The greatest growth in the nuclear industry has been in the development of nuclear power plants. Today more than 400 nuclear reactors generate electricity around the world. Almost one-quarter of them can be found in one country—the United States—and most of the rest are located in a relatively small number of countries—most notably France, Japan, Russia, South Korea, India, Canada, the United Kingdom, and China. Clearly, nuclear power is an important branch of nuclear engineering. It includes a number of specialties.

Reactor Physics and Radiation Transport

Nuclear engineers analyze the complex physical and radiation transport phenomena occurring within and external to a nuclear reactor. Working closely with engineers and scientists from other fields, they model complex phenomena such as heat transfer, fluid flow, chemical reactions, and materials response. Modeling and simulation play a dominant role in this area.

Reactor Thermal Hydraulics and Heat Transfer

The energy released by fission is carried by the reactor coolant out of the reactor core and is used to create steam to turn a turbine and generate electricity. Nuclear engineers work with mechanical engineers to determine the heat transfer and coolant flow within the reactor, typically using complex simulation codes to model the combined nuclear and thermal hydraulic phenomena.

Core Design

Given specifications for a particular power plant, the core designer employs modeling and simulation tools to arrive at an optimal design that satisfies the performance specifications and also meets the regulatory criteria.

Safety Analysis

This area includes both deterministic and probabilistic safety analyses (the latter also known as probabilistic risk assessments, or PRAs). A deterministic safety analysis evaluates the response of the reactor plant to normal operating conditions, anticipated abnormal operating conditions, and postulated accident situations. The analysis involves modeling of complex phenomena including neutron transport, thermal hydraulics, heat transfer, structural analysis, and radiation effects on material properties. Advanced modeling and simulation tools on large mainframe computers are needed to carry out these simulations.

PRAs estimate the risk associated with various accident scenarios by estimating the probability of the event's occurring and then the probable consequences of the event. Typically PRAs are performed by constructing "event trees," which follow an accident progressing from some assumed initiating event, and "fault trees," which work backward from an assumed failure in order to determine the probability of that failure's occurring. Since the Fukushima accident of 2011, heavy emphasis has been placed on safety analysis, both deterministic and probabilistic.

Fuel Management

Fuel management involves specifying, procuring, and managing fuel throughout its reactor lifetime and beyond. Also known as "core follow," this involves the optimal placement ("shuffling") of new fuel and old fuel in the reactor during refueling stages.

Navy Nuclear Propulsion

The reactors used for submarines and surface ships are similar to commercial pressurized-water reactors, except they are smaller and more rugged in order to withstand battle conditions such as torpedoes or depth charges or extreme loads such as the catapulting of jets off a carrier deck. Many of the functions carried out by nuclear engineers in working for the navy are similar to functions for electricity generation. Indeed, a large number of "navy nukes" find employment in the commercial nuclear power industry after leaving the navy.

Fusion Energy and Plasma Physics

Nuclear fusion is a potential energy source with a wide range of applications. The fusion process is the opposite of the fission process, as it proceeds by combining two light nuclei in an ionized gaseous state, or plasma, to form a heavier nucleus that has less mass than the two original nuclei because of a release of energy. Nuclear fusion powers the universe, being the source of energy in the Sun and the stars. The energy from a fusion reaction can be released in a variety of forms, including charged particles, electromagnetic radiation, and neutrons, but the key challenge for nuclear engineers and plasma physicists is to control the reaction, similar to controlling the fission reaction in a nuclear reactor. If a fusion reactor were to be built, water could be used as fuel, thus providing an inexhaustible source of energy for society. This great potential has motivated initiatives such as ITER (International Thermonuclear Experimental Reactor), a multinational program to build a practical fusion power plant. Nuclear engineers work with plasma physicists to design and analyze fusion power plants as well as to understand the physics of plasmas and their applications.

Nuclear Weapons

Fission weapons (atomic bombs), fusion weapons (hydrogen bombs), and combination fission-fusion weapons make up the world's nuclear arsenal. Nuclear engineers employed in weapons programs engage in such diverse activities as research, development, design, fabrication, production, testing, maintenance, and surveillance of a large array of nuclear weapons systems. Since a nuclear weapon is a complex engineering system, scientists and engineers from many fields are needed to build it. Specific training and education courses on nuclear weapons will not be found in a nuclear engineering curriculum, as the field is highly classified and subject to security rules.

Radioisotopes

More than 2,000 radioactive isotopes are produced in nuclear reactors, and nuclear engineers are involved in both their production and their use. The production, packaging, and application of many of these isotopes have become a large industry. They are used in heart pacemakers, medical research, sterilization of medical instruments, industrial tracers, X-ray equipment, curing of plastics, and preservation of food and also as an energy source in electric generators. Perhaps the most important use of radioisotopes is in the field of medicine. One notable example is molybdenum-99, a fission product that decays to the short-lived gamma-ray-emitting isotope technetium-99m, a nuclear isomer used in various imaging applications in medicine.

Nuclear-waste Management

Nuclear wastes can be classified in two groups, low-level and high-level. Low-level wastes come from nuclear power facilities, hospitals, and research institutions and include such items as contaminated clothing, wiping rags, tools, test tubes, needles, and other medical research materials. Low-level waste is packaged in leak-proof containers and placed in earth-covered trenches at a low-level-waste disposal site. High-level wastes are highly radioactive and derive from spent fuel elements and from weapons programs. In theory, these wastes are to be disposed of in permanent facilities deep underground, but in fact no country with a civilian or military nuclear program has begun to do so. In the United States, for instance, high-level waste from the nuclear weapons program has been stored since 1999 at the Waste Isolation Pilot Plant (WIPP) in New Mexico, while construction work has been started and stopped on a proposed permanent repository beneath Yucca Mountain in Nevada. Nuclear engineers are involved in the design of permanent repositories, which includes analyses of the effects of radiation and decay heat on containers and geological formations.

Nuclear Materials

Materials used in nuclear reactors are subjected to high temperatures and radiation, and these extreme conditions cause a degradation of the materials' properties. Nuclear engineers study the effects of radiation on materials in order to develop new radiation-resistant materials or to determine when degraded material should be replaced. Objects exposed to high radiation levels include nuclear fuel, internal components of a reactor, containers for storage of high-level nuclear waste, and the materials that make up the nuclear waste itself. By studying and exploiting the fundamental changes that occur in materials as a result of irradiation, it is possible to develop new materials that would not be obtainable through conventional methods.

Radiation Measurements

Nuclear engineers working in the area of radiation measurements develop advanced detection and measurement systems that can also be used to improve imaging technologies. Their work includes detector design, fabrication of new detectors and analysis of their performance, measurements of fundamental atomic and nuclear phenomena that are needed for nuclear reactor analysis, development of new algorithms and methods for detector systems, neutron activation analysis, nondestructive testing, and evaluation of components using penetrating radiation. Nuclear engineers also develop and apply advanced radiation-detection technologies to combat proliferation of nuclear weapons and guard against nuclear terrorism, for example, in the development of systems capable of detecting nuclear materials in shipping containers.

Medical and Health Physics

Medical physicists and radiation oncologists may employ radiation for diagnosis and therapy, whereas health physicists deal with the effects on humans of ionizing radiation encountered, for example, through occupational exposure. In all cases, nuclear engineers may be involved with analyzing the transport of radiation within a human being and in assessing the biological effect of radiation on healthy as well as diseased tissue.

Computation in Nuclear Engineering

From the start of nuclear energy development, computation was a key technology for nuclear reactor design and operation. The computer development is one of the largest development among of scientific technologies. And the nuclear engineering computation becomes more and more accurate and sophisticated along with this development.

Current Computation in Nuclear Engineering

Areas of Numerical Analyses in Nuclear Engineering

Thise topic discusses mainly on nuclear reactors, since we have only a limited number of pages. For each area we have many sub-areas to be analyzed by calculation, for example, the fuel cycle, which covers mining, enrichment, fuel fabrication, reprocessing, waste disposal, and fuel transportation. On the other hand some areas may be overlapped, for example, the cost is often investigated with optimization. Required computation method usually differs in different areas.

- Cross section evaluations,
- Reactor physics analysis (mainly neutron transport),
- Thermal aspects including fluid behaviors,
- Instrumentations and control,
- Stability, operations research,
- Stress analyses of solid components,
- Diffusion of materials especially in fuel elements,
- Fuel cycle including wastes disposal,

- Safety and accidents,

- Seculity,

- Environmental effects of radioactive materials from facilities, environment dynamics,

- Effects of irradiations on materials and human body,

- Designs of reactor system and elements,

- Costs,

- Optimization.

Procedures for Solution

The general procedures for solution is shown in figure. Usually problem definition and formulation are combined and called as problem formulation.

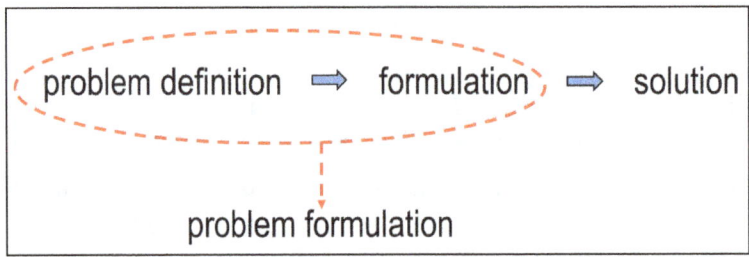

Procedures for Solution.

Albert Einstein said "The formulation of a problem is often far more essential than its solution." This is true especially for physics. However, for the present nuclear engineering the problem is already presented and our jobs are usually solutions.

Many calculation methods are available for solving these problems. The problems for neutron transport are given by differential and integral equations. Usually these equations are transformed to matrix equations by employing discretization of continuous variables such as spatial coordinates, time, energy and angle. Electric computer is powerful for solving these equations. The neutron transport is also solved by Monte Carlo method, where discretization of continuous variables is not necessary. Since each method has its original merits and demerits, we should be careful for choosing proper method for each problem.

For safety analysis, we usually employ probabilistic risk analysis (PRA) which include event tree analysis and fault tree analysis. These analyses are entirely different from the neutron transport analysis. For this kind of problem, difficult part is not solving equation, but obtaining proper inputs.

We meet often optimization and stability problems. These analyses are also very different from the neutron transport analysis. Especially we have so many unique optimization methods from analytical to heuristic. We need to be familiar to these methods.

Figure shows the only one direction from the problem formulation to solution, but after obtaining a solution we should check the whole procedure and change the problem formulation and solution, if necessary.

Limits of Analysis

When more contents are included in the analysis of problem, we can get more informative results. More general results can be obtained by treating wider problem region. More accurate results can be obtained by changing the assumption employed for the calculation to less approximated. More accurate results are obtained when number of calculation meshes is increased and convergence criteria are changed to be smaller. These improvements of analysis increase the computation size. The computations are usually performed by using computer. Both of the accuracy of results and problem size are limited by computer performance as shown in figure.

$$(\text{Accuracy of Result}) \times (\text{Problem Size}) < (\text{Computer Perfoemance})$$

Increase of Computer Performance

The computer performance is increasing according to Moor's law "the number of transistors in a dense integrated circuit doubles approximately every two years" for many years as shown in figure.

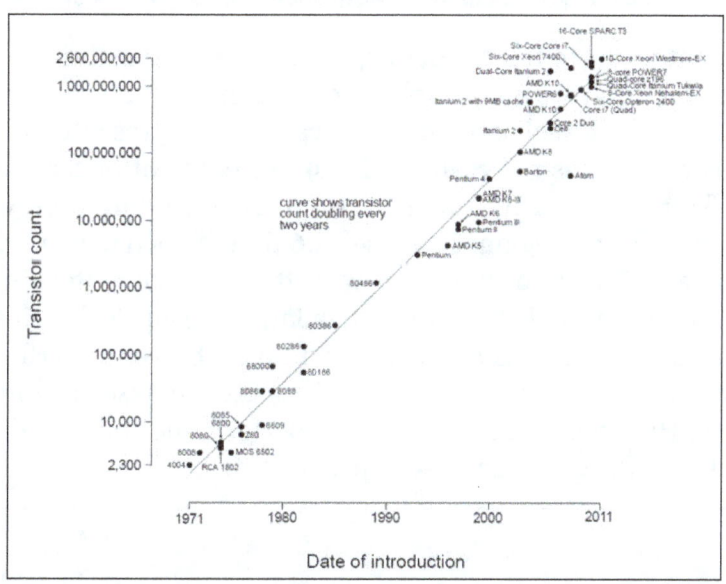

Increase of computer performance: microprocessor transistor counts 1971-2011.

The change of this computer performance enables the larger computation in the later years. One example is shown in figure, which is in-core fuel management problem.

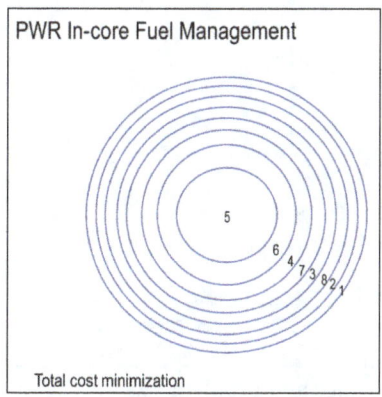

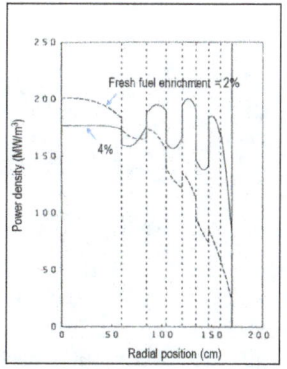

LBE Cooled Fast Reactor In-core Fuel Management

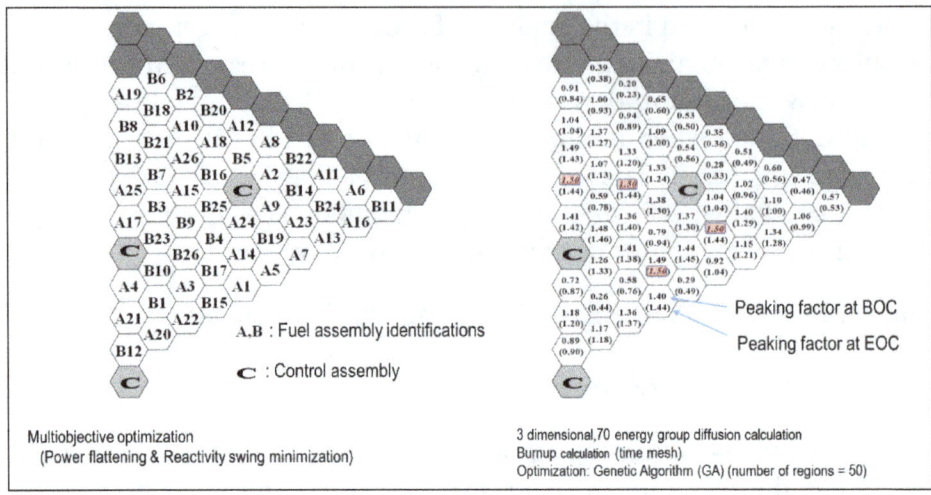

Effects of increase of computer performance on actual reactor analyses.

In-core fuel management needs to treat a lot of space meshes for treating fine fuel assembly positions and a lot of cases for different fuel shuffling scheme. In 1973 it was impossible to treat actual fuel shuffling scheme of each fuel assembly. The calculation presented in figure above treated cylindrically divided core which is not realistic. In 1999 treatment of each fuel assembly became possible as shown in figure above. The total mesh numbers including energy groups increased by more than 104 and the number of regions increased 5.6 times which increased the number of shuffling patterns so drastically. The used computers were the large computers in university computer center: University of California, Berkeley for the calculation (a) and Tokyo Institute of Technology for the calculation (b). Both machines had the world highest level of performance at each time. This situation is shown in figure. The treatable problem size increased so much by the increase of computer performance. However, not only computer performance but developments of analytical method also contributed to the great increase of problem size.

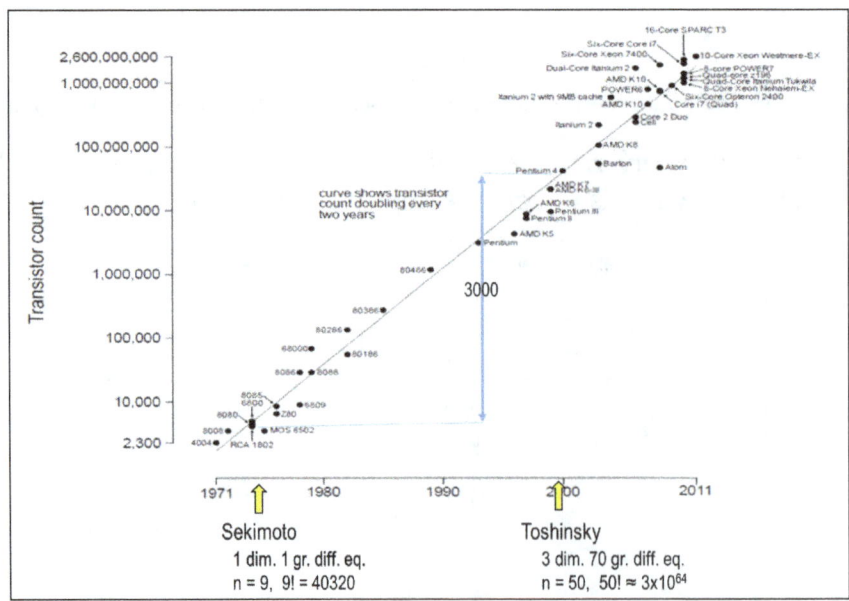

Increase of computer performance and problem size with year.

Future Computation in Nuclear Engineering

Computation in Future Nuclear Engineering

The computer performance will continue to increase in the future. Several innovative computers such as quantum computers are expected to be put to practical use in the near future. It will be able to handle larger and more complicated systems faster and more accurately. In the future of nuclear power, the role of computation should increase more and more. One case of the computation in future nuclear engineering is for a new field of innovative nuclear engineering under development. In another case, the development of the computer enables calculation that was impossible to date. Here in this topic an interesting example of computation is shown for each case: innovative nuclear reactors for the fast case and accident managements for the second case.

Computation for Innovative Nuclear Reactor System

The computation technique will be improved even for the basic problems of conventional reactors. However, developments of computation technique are considered much more for innovative reactors. Some of the important innovative reactors are shown in table with their unique characteristics.

Rctor types	Unique characteristics
Fast Reactor	Hard neutron spectrum Re-criticality accidents
High Temperature Gas Cooled Reactor	Neutron self shielding for double heterogeneity depressurizing accident
Molten Salt Reactor	Liquid fuel (FP flow, on-line reprocessing)
Accelerator Driven System	Sub-criticality Unisotropic neutron field Hard neutron spectrum
Fusion Reactor Blanket	Sub-criticality Unisotropic neutron field Hard neutron spectrum

Each reactor type in this table contains many different subtypes. The fast reactor has many options for different fuel: oxide, metal, nitride, et al., and for different coolants: sodium, lead (or lead bismuth), gas, et al. The high temperature gas cooled reactor has typical two options: block-type fuel reactor and pebble bed reactor. The molten salt reactor has two types: thermal thorium reactor and fast reactor. The accelerator driven system uses solid core or liquid core. The fusion reactor blanket has two options; simple tritium breeding blanket and complicated fusion-fission hybrid blanket. It has also many options for different plasma core design, and many subtypes of design like the fast reactor. Many innovative reactors other than those listed in this table have been proposed.

 The unique characteristics shown in table are only those related to reactor physics. The FP flow for the molten salt reactor includes both delayed neutron precursor and high neutron absorbers like Xe-135. They affect greatly the reactor physics characteristics. Proper computation method should be developed to treat these characteristics shown in table.

Computation for Accident Management

Experiences of Fukushima-Daiichi accident provided a lot of suggestions for protecting and mitigating such a severe accident. Quick proper response to the accident is one of the most important requirements. The director of reactor has to know all details of the reactor and should make an appropriate judgment on what to do when problems arise. However, this ability has been greatly exceeding that of ordinary human beings. Furthermore, at the accident all of necessary data cannot be obtained in most cases, and human error easily occur. Mental stress of the director and operators become too strong.

Computer can make a more calm judgment as compared to humans. It can make a proper judgment instantaneously from a lot of information. In case of an accident it is conceivable that the computer makes a decision on behalf of human beings.

All of the reactor information must be given to the computer. In an accident, some information may be impossible to be given due to disconnection of the signal cable or some other reasons. We have to build a system so that this kind of thing does not happen, but even if it happens, the computer system must make appropriate judgments with available information.

Rapid judgment is inevitable though the problem is so large and complicated. The artificial intelligence (AI) seems very efficient tool for this problem. The AI programs for games such as chess and go-game has been developed drastically for recent years. Some of the technology of game programs may be employed in our program.

To err is human, and the responsibility for a nuclear accident is too heavy for the humans. The author considers humans should not make such judgements. However, according to usual human rules, it may appear that there is a doubt about the judgment of the computer. In such a case, it may become necessary to make a system directed by humans that can communicate with the computer.

Materials for Nuclear Engineering

The knowledge of thermophysical and nuclear properties of materials is essential for designing nuclear power plants. In general, there are two basic types of materials, that are used in nuclear power plant.

- Materials having specific nuclear properties - These materials must fulfill very specific requirements that originates especially in interactions of atomic nuclei. This class corresponds to the nuclear fuels, neutron absorbing materials or alloys of low neutron capture cross-sections. For these materials proper nuclear properties are a priority and the best chemical state (atomic properties) could be selected (e.g., boron as a neutron absorber can be used in water solution as boric acid or as boron carbide – B_4C in control rods).

- Standard engineering materials - These materials does not differ from material used in other engineering branches. It must be added nuclear engineering puts higher requirements on quality and reliability of all materials than in other engineering branches. This class corresponds, for example, to alloys, such as structural steels, stainless steels, aluminum alloys, etc. Few specific alloys have been developed for particular applications, such as Zr alloys in water reactors, which belong also to the materials having specific nuclear properties.

Materials essential for designing nuclear power plants can be divided into the following groups:

- Nuclear Fuels - Nuclear fuel is generally any material that can be 'burned' by nuclear fission to derive nuclear energy. Common nuclear reactors use an enriched uranium and plutonium as a fuel. Most of PWRs use the uranium fuel, which is in the form of uranium dioxide, but other fuels and matrices are developed.

- Neutron Moderators - The moderator, which is of importance in thermal reactors, is used to moderate, that is, to slow down, neutrons from fission to thermal energies. Commonly used moderators include regular (light) water (roughly 75% of the world's reactors), solid graphite (20% of reactors) and heavy water (5% of reactors). Beryllium and beryllium oxide (BeO) have been used occasionally, but they are very costly.

- Neutron Absorbers - The materials that absorb neutrons are used in reactor core in the following three cases:

  ○ In control rods, that are an important safety system of nuclear reactors.

  ○ As burnable absorbers, which can be dispersed uniformly in fuel or placed in certain sections.

  ○ As additives to a moderator for compensation of an excess reactivity.

- Coolants - In nuclear power plant, water and steam are a common fluids used for heat exchange in the primary circuit (from surface of fuel rods to the coolant flow) and in the secondary circuit. But many other materials can be used for this purpose. In power reactors, carbon dioxide, heavy water, helium of liquid metals can be used.

- Structural Materials - Many various materials are used in the designs of nuclear reactors. Some materials must have lower neutron capture cross-section, especially inside a reactor core, where fission chain reaction takes place. On the other hand there are many materials, that must have higher capture cross-sections.

- Radiation Shielding - Radiation shielding usually consist of barriers of lead, concrete or water. There are many many materials, which can be used for radiation shielding, but there are many many situations in radiation protection. It highly depends on the type of radiation to be shielded, its energy and many other parametres. For example, even depleted uranium can be used as a good protection from gamma radiation, but on the other hand uranium is absolutely inappropriate shielding of neutron radiation.

Radioactivity

Radioactivity is property exhibited by certain types of matter of emitting energy and subatomic particles spontaneously. It is, in essence, an attribute of individual atomic nuclei.

An unstable nucleus will decompose spontaneously, or decay, into a more stable configuration but will do so only in a few specific ways by emitting certain particles or certain forms of electromagnetic energy. Radioactive decay is a property of several naturally occurring elements as well as of artificially produced isotopes of the elements. The rate at which a radioactive element decays is expressed in terms of its half-life; i.e., the time required for one-half of any given quantity of the isotope to decay. Half-lives range from more than 1,000,000,000 years for some nuclei to less than 10^{-9} second. The product of a radioactive decay process—called the daughter of the parent isotope—may itself be unstable, in which case it, too, will decay. The process continues until a stable nuclide has been formed.

Nature of Radioactive Emissions

The emissions of the most common forms of spontaneous radioactive decay are the alpha (α) particle, the beta (β) particle, the gamma (γ) ray, and the neutrino. The alpha particle is actually the nucleus of a helium-4 atom, with two positive charges ^4_2He. Such charged atoms are called ions. The neutral helium atom has two electrons outside its nucleus balancing these two charges. Beta particles may be negatively charged (beta minus, symbol e^-), or positively charged (beta plus, symbol e^+). The beta minus [β^-] particle is actually an electron created in the nucleus during beta decay without any relationship to the orbital electron cloud of the atom.

The beta plus particle, also called the positron, is the antiparticle of the electron; when brought together, two such particles will mutually annihilate each other. Gamma rays are electromagnetic radiations such as radio waves, light, and X-rays. Beta radioactivity also produces the neutrino and antineutrino, particles that have no charge and very little mass, symbolized by v and $\bar{v}$ respectively.

In the less common forms of radioactivity, fission fragments, neutrons, or protons may be emitted. Fission fragments are themselves complex nuclei with usually between one-third and two-thirds the charge Z and mass A of the parent nucleus. Neutrons and protons are, of course, the basic building blocks of complex nuclei, having approximately unit mass on the atomic scale and having zero charge or unit positive charge, respectively. The neutron cannot long exist in the free state. It is rapidly captured by nuclei in matter; otherwise, in free space it will undergo beta-minus decay to a proton, an electron, and an antineutrino with a half-life of 12.8 minutes. The proton is the nucleus of ordinary hydrogen and is stable.

Types of Radioactivity

The early work on natural radioactivity associated with uranium and thorium ores identified two distinct types of radioactivity: alpha and beta decay.

Alpha Decay

In alpha decay, an energetic helium ion (alpha particle) is ejected, leaving a daughter nucleus of atomic number two less than the parent and of atomic mass number four less than the parent. An example is the decay (symbolized by an arrow) of the abundant isotope of uranium, 238U, to a thorium daughter plus an alpha particle:

$$^{238}_{92}\text{U} \rightarrow {}^{234}_{90}\text{Th} + {}^4_2\text{He}$$

$$Q_\alpha = 4.268\,\text{MeV}$$

$$t_{1/2} = 4051 \times 10^9 \text{ years}$$

Given for this and subsequent reactions are the energy released (Q) in millions of electron volts (MeV) and the half-life ($t_{1/2}$). It should be noted that in alpha decays the charges, or number of protons, shown in subscript are in balance on both sides of the arrow, as are the atomic masses, shown in superscript.

Beta-minus Decay

In beta-minus decay, an energetic negative electron is emitted, producing a daughter nucleus of one higher atomic number and the same mass number. An example is the decay of the uranium daughter product thorium-234 into protactinium-234:

$$^{234}_{90}\text{U} \rightarrow \, ^{234}_{91}\text{Pa} + e^- + \bar{v}$$

$$Q_{\beta+} = .263 \, \text{MeV}$$

$$t_{1/2} = 24.1 \, \text{days}$$

In the above reaction for beta decay $\bar{v}$ represents the antineutrino. Here, the number of protons is increased by one in the reaction, but the total charge remains the same, because an electron, with negative charge, is also created.

Gamma Decay

A third type of radiation, gamma radiation, usually accompanies alpha or beta decay. Gamma rays are photons and are without rest mass or charge. Alpha or beta decay may simply proceed directly to the ground (lowest energy) state of the daughter nucleus without gamma emission, but the decay may also proceed wholly or partly to higher energy states (excited states) of the daughter. In the latter case, gamma emission may occur as the excited states transform to lower energy states of the same nucleus. (Alternatively to gamma emission, an excited nucleus may transform to a lower energy state by ejecting an electron from the cloud surrounding the nucleus. This orbital electron ejection is known as internal conversion and gives rise to an energetic electron and often an X-ray as the atomic cloud fills in the empty orbital of the ejected electron. The ratio of internal conversion to the alternative gamma emission is called the internal-conversion coefficient).

Isomeric Transitions

There is a wide range of rates of half-lives for the gamma-emission process. Usually dipole transitions, in which the gamma ray carries off one $\hbar$ unit of angular momentum, are fast, less than nanoseconds (one nanosecond equals 10^{-9} second). The law of conservation of angular momentum requires that the sum of angular momenta of the radiation and daughter nucleus is equal to the angular momentum (spin) of the parent. If the spins of initial and final states differ by more than one, dipole radiation is forbidden, and gamma emission must proceed more slowly by a higher multipole (quadrupole, octupole, etc.) gamma transition. If the gamma-emission half-life exceeds about one nanosecond, the excited nucleus is said to be in a metastable, or isomeric, state (the names for a long-lived excited state), and it is customary to classify the decay as another type of radioactivity, an isomeric transition. An example of isomerism is found in the protactinium-234 nucleus of the uranium-238 decay chain:

$$^{234m}_{91}\text{Pa} \rightarrow \, ^{234}_{91}\text{Pa} + \gamma$$

$$Q_\gamma = 0.0698 \, \text{MeV}$$

$$t_{1/2} = 1.17 \, \text{min}$$

The letter M following the mass number stands for metastable and indicates a nuclear isomer.

Beta-plus Decay

During the 1930s new types of radioactivity were found among the artificial products of nuclear reactions: beta-plus decay, or positron emission, and electron capture. In beta-plus decay an energetic positron is created and emitted, along with a neutrino, and the nucleus transforms to a daughter, lower by one in atomic number and the same in mass number. For instance, carbon-11 ($Z = 6$) decays to boron-11 ($Z = 5$), plus one positron and one neutrino:

$$^{11}_{6}C \rightarrow {}^{11}_{5}B + e^{+} + v$$
$$Q_{\beta+} = 0.97\,MeV$$
$$t_{1/2} = 20.4\,min$$

Electron Capture

Electron capture (EC) is a process in which decay follows the capture by the nucleus of an orbital electron. It is similar to positron decay in that the nucleus transforms to a daughter of one lower atomic number. It differs in that an orbital electron from the cloud is captured by the nucleus with subsequent emission of an atomic X-ray as the orbital vacancy is filled by an electron from the cloud about the nucleus. An example is the nucleus of beryllium-7 capturing one of its inner electrons to give lithium-7:

$$^{7}_{4}Be \rightarrow {}^{7}_{3}Li + v$$
$$QEC = 0.8616\,MeV$$
$$t_{1/2} = 53\,days$$

The main features of radioactive decay of a nuclear species are often displayed in a decay scheme. Figure shows the decay scheme of beryllium-7. Indicated are the half-life of the parent and that of the excited daughter state, as well as its energy 0.4774 MeV. The spins and parities of all three states are provided on the upper left-hand side of the level. The multipolarity of the gamma ray (magnetic dipole, M_1, plus 0.005 percent electric quadrupole, E2) is indicated above the vertical arrow symbolizing the gamma transition. The slanted arrows symbolize the electron-capture decay with labels giving the percentage of decay directly to ground state (89.7 percent) and the percentage of EC decay going via the excited state (10.3 percent). The bold-face numbers following the percentages are so-called log ft values, to be encountered below in connection with beta-decay rates. The overall energy release, Q_{EC}. The Q_{EC} is necessarily a calculated value because there is no general practical means of measuring the neutrino energies accompanying EC decay. With a few electron-capturing nuclides, it has been possible to measure directly the decay energy by measurement of a rare process called inner bremsstrahlung (braking radiation). In this process the energy release is shared between the neutrino and a gamma ray. The measured distribution of gamma-ray energies indicates the total energy release. Usually there is so much ordinary gamma radiation with radioactive decay that the inner bremsstrahlung is unobservable.

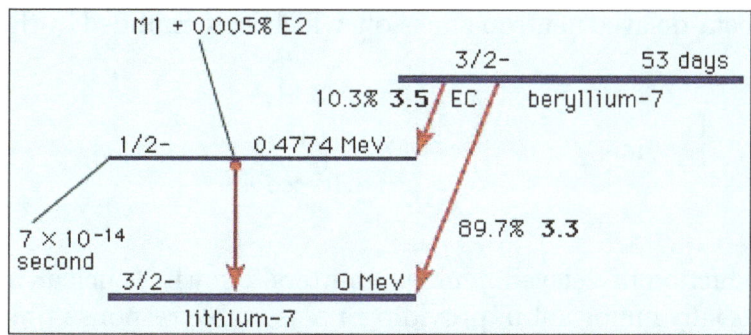

Radioactive decay of beryllium-7 to lithium-7 by electron capture.

Spontaneous Fission

Yet another type of radioactivity is spontaneous fission. In this process the nucleus splits into two fragment nuclei of roughly half the mass of the parent. This process is only barely detectable in competition with the more prevalent alpha decay for uranium, but for some of the heaviest artificial nuclei, such as fermium-256, spontaneous fission becomes the predominant mode of radioactive decay. Kinetic-energy releases from 150 to 200 MeV may occur as the fragments are accelerated apart by the large electrical repulsion between their nuclear charges. The reaction is as follows:

$$^{256}_{100}\text{Fm} \rightarrow {}^{140}_{54}\text{Xe} + {}^{112}_{46}\text{Pd} + 4\text{n} + \text{other fission product}$$

$$t_{1/2} = 2.7 \text{hours}$$

$$Q = 150 - 200 \text{MeV}$$

Only one of several product sets is shown. A few neutrons are always emitted in fission of this isotope, a feature essential to chain reactions. Spontaneous fission is the process involved in nuclear reactors. Induced fisson is a property of uranium-235, plutonium-239, and other isotopes to undergo fission after absorption of a slow neutron. Other than the requirement of a neutron capture to initiate it, induced fission is quite similar to spontaneous fission regarding total energy release, numbers of secondary neutrons, and so on.

Proton Radioactivity

Proton radioactivity, discovered in 1970, is exhibited by an excited isomeric state of cobalt-53, ^{53m}Co, 1.5 percent of which emits protons:

$$^{53m}_{27}\text{Co} \underset{98.5\%}{\overset{1.5\%}{\diagdown}} \begin{array}{l} {}^{52}_{26}\text{Fe} + p \qquad\qquad Q_p = 1.57 \text{ MeV} \\[2ex] {}^{53}_{26}\text{Fe} + e^+ + \nu \qquad t_{1/2} = 0.243 \text{ sec} \end{array}$$

Special Beta-decay Processes

In addition to the above types of radioactivity, there is a special class of rare beta-decay processes that gives rise to heavy-particle emission. In these processes the beta decay partly goes to a high excited state of the daughter nucleus, and this state rapidly emits a heavy particle.

One such process is beta-delayed neutron emission, which is exemplified by the following reaction:

$$\begin{aligned}
{}^{17}_{7}\text{N} &\longrightarrow {}^{17}_{8}\overset{*}{\text{O}} + e^- + \bar{\nu} & Q_{\beta-} &= 8.68 \text{ MeV} \\
&\phantom{\longrightarrow {}^{17}_{8}}\hookrightarrow {}^{16}_{8}\text{O} + n & E_{max\,n} &= 1.81 \text{ MeV} \\
& & t_{1/2} &= 4.16 \text{ sec}
\end{aligned}$$

There is a small production of delayed neutron emitters following nuclear fission, and these radioactivities are especially important in providing a reasonable response time to allow control of nuclear fission reactors by mechanically moved control rods.

Among the positron emitters in the light-element region, a number beta decay partly to excited states that are unstable with respect to emission of an alpha particle. Thus, these species exhibit alpha radiation with the half-life of the beta emission. Both the positron decay from boron-8 and electron decay from lithium-8 are beta-delayed alpha emission, because ground as well as excited states of beryllium-8 are unstable with respect to breakup into two alpha particles. Another example, sodium-20 (^{20}Na) to give successively neon-20 (^{20}Ne; the asterisk again indicating the short-lived intermediate state) and finally oxygen-16 is listed below:

$$\begin{aligned}
{}^{20}_{11}\text{Na} &\longrightarrow {}^{20}_{10}\overset{*}{\text{Ne}} + e^+ + \nu & Q_{\beta+} &= 13.0 \text{ MeV} \\
&\phantom{\longrightarrow {}^{20}_{10}}\hookrightarrow {}^{16}_{8}\text{O} + \alpha & E_{max\,\alpha} &= 4.44 \text{ MeV} \\
& & t_{1/2} &= 0.39 \text{ sec}
\end{aligned}$$

In a few cases, positron decay leads to an excited nuclear state not able to bind a proton. In these cases, proton radiation appears with the half-life of the beta transition. The combination of high positron-decay energy and low proton-binding energy in the daughter ground state is required. In the example given below, tellurium-111 (^{111}Te) yields antimony-111 (^{111}Sb) and then tin-110 (^{110}Sn) successively:

$$\begin{aligned}
{}^{111}_{52}\text{Te} &\longrightarrow {}^{111}_{51}\overset{*}{\text{Sb}} + e^+ + \nu & Q_{\beta+} &\text{ uncertain} \\
&\phantom{\longrightarrow {}^{111}_{51}}\hookrightarrow {}^{110}_{50}\text{Sn} + p & E_{max\,p} &= 3.7 \text{ MeV} \\
& & t_{1/2} &= 19.5 \text{ sec}
\end{aligned}$$

Heavy-ion Radioactivity

In 1980 A. Sandulescu, D.N. Poenaru, and W. Greiner described calculations indicating the possibility of a new type of decay of heavy nuclei intermediate between alpha decay and spontaneous fission. The first observation of heavy-ion radioactivity was that of a 30-MeV, carbon-14 emission from radium-223 by H.J. Rose and G.A. Jones in 1984. The ratio of carbon-14 decay to alpha decay is about 5×10^{-10}. Observations also have been made of carbon-14 from radium-222, radium-224, and radium-226, as well as neon-24 from thorium-230, protactinium-231, and uranium-232. Such heavy-ion radioactivity, like alpha decay and spontaneous fission, involves quantum-mechanical tunneling through the potential-energy barrier. Shell effects play a major role in this phenomenon, and in all cases observed to date the heavy partner of carbon-14 or neon-24 is close to doubly magic lead-208.

Occurrence of Radioactivity

Some species of radioactivity occur naturally on Earth. A few species have half-lives comparable to the age of the elements (about 6×10^9 years), so that they have not decayed away after their formation in stars. Notable among these are uranium-238, uranium-235, and thorium-232. Also, there is potassium-40, the chief source of irradiation of the body through its presence in potassium of tissue. Of lesser significance are the beta emitters vanadium-50, rubidium-87, indium-115, tellurium-123, lanthanum-138, lutetium-176, and rhenium-187, and the alpha emitters cerium-142, neodymium-144, samarium-147, gadolinium-152, dysprosium-156, hafnium-174, platinum-190, and lead-204. Besides these approximately 109-year species, there are the shorter-lived daughter activities fed by one or another of the above species; e.g., by various nuclei of the elements between lead (Z = 82) and thorium (Z = 90).

Another category of natural radioactivity includes species produced in the upper atmosphere by cosmic ray bombardment. Notable are 5,720-year carbon-14 and 12.3-year tritium (hydrogen-3), 53-day beryllium-7, and 2,700,000-year beryllium-10. Meteorites are found to contain additional small amounts of radioactivity, the result of cosmic ray bombardments during their history outside the Earth's atmospheric shield. Activities as short-lived as 35-day argon-37 have been measured in fresh falls of meteorites. Nuclear explosions since 1945 have injected additional radioactivities into the environment, consisting of both nuclear fission products and secondary products formed by the action of neutrons from nuclear weapons on surrounding matter.

The fission products encompass most of the known beta emitters in the mass region 75–160. They are formed in varying yields, rising to maxima of about 7 percent per fission in the mass region 92–102 (light peak of the fission yield versus atomic mass curve) and 134–144 (heavy peak). Two kinds of delayed hazards caused by radioactivity are recognized. First, the general radiation level is raised by fallout settling to Earth. Protection can be provided by concrete or earth shielding until the activity has decayed to a sufficiently low level. Second, ingestion or inhalation of even low levels of certain radioactive species can pose a special hazard, depending on the half-life, nature of radiations, and chemical behaviour within the body.

Nuclear reactors also produce fission products but under conditions in which the activities may be contained. Containment and waste-disposal practices should keep the activities confined and eliminate the possibility of leaching into groundwaters for times that are long compared to the half-lives. A great advantage of thermonuclear fusion power over fission power, if it can be practically realized, is not only that its fuel reserves, heavy hydrogen and lithium, are vastly greater than uranium, but also that the generation of radioactive fission product wastes can be largely avoided. In this connection, it may be noted that a major source of heat in the interior of both the Earth and the Moon is provided by radioactive decay. Theories about the formation and evolution of the Earth, Moon, and other planets must take into account these large heat production sources.

Desired radioactivities other than natural activities and fission products may be produced either by irradiation of certain selected target materials by reactor neutrons or by charged particle beams or gamma ray beams of accelerators.

Energetics and Kinetics of Radioactivity

Energy Release in Radioactive Transitions

Consideration of the energy release of various radioactive transitions leads to the fundamental question of nuclear binding energies and stabilities. A much-used method of displaying nuclear-stability relationships is an isotope chart, those positions on the same horizontal row corresponding to a given proton number (Z) and those on the same vertical column to a given neutron number (N). The irregular bold line surrounds the region of presently known nuclei. The area encompassed by this is often referred to as the valley of stability because the chart may be considered a map of a binding energy surface, the lowest areas of which are the most stable. The most tightly bound nuclei of all are the abundant iron and nickel isotopes. Near the region of the valley containing the heaviest nuclei (largest mass number A; i.e., largest number of nucleons, N + Z), the processes of alpha decay and spontaneous fission are most prevalent; both these processes relieve the energetically unfavourable concentration of positive charge in the heavy nuclei.

Along the region that borders on the valley of stability on the upper left-hand side are the positron-emitting and electron-capturing radioactive nuclei, with the energy release and decay rates increasing the farther away the nucleus is from the stability line. Along the lower right-hand border region, beta-minus decay is the predominant process, with energy release and decay rates increasing the farther the nucleus is from the stability line.

The grid lines of the graph are at the nucleon numbers corresponding to extra stability, the "magic numbers". The circles labeled "deformed regions" enclose regions in which nuclei should exhibit cigar shapes; elsewhere the nuclei are spherical. Outside the dashed lines nuclei would be unbound with respect to neutron or proton loss and would be exceedingly short-lived (less than 10^{-19} second).

Calculation and Measurement of Energy

By the method of closed energy cycles, it is possible to use measured radioactive-energy-release (Q) values for alpha and beta decay to calculate the energy release for unmeasured transitions. An illustration is provided by the cycle of four nuclei below:

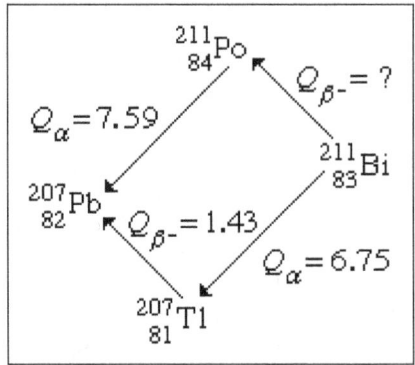

In this cycle, energies from two of the alpha decays and one beta decay are measurable. The unmeasured beta-decay energy for bismuth-211, $Q_\beta-(B_i)$, is readily calculated because conservation of energy requires the sum of Q values around the cycle to be zero. Thus, $Q_\beta-(B_i) + 7.59 - 1.43$

– 6.75 = 0. Solving this equation gives Q_{β^-}(Bi) = 0.59 MeV. This calculation by closed energy cycles can be extended from stable lead-207 back up the chain of alpha and beta decays to its natural precursor uranium-235 and beyond. In this manner the nuclear binding energies of a series of nuclei can be linked together. Because alpha decay decreases the mass number A by 4, and beta decay does not change A, closed α–β-cycle calculations based on lead-207 can link up only those nuclei with mass numbers of the general type A = 4n + 3, in which n is an integer. Another, the 4n series, has as its natural precursor thorium-232 and its stable end product lead-208. Another, the 4n + 2 series, has uranium-238 as its natural precursor and lead-206 as its end product.

In early research on natural radioactivity, the classification of isotopes into the series cited above was of great significance because they were identified and studied as families. Newly discovered radio activities were given symbols relating them to the family and order of occurrence therein. Thus, thorium-234 was known as UX1, the isomers of protactinium-234 as UX_2 and UZ, uranium-234 as UII, and so forth. These original symbols and names are occasionally encountered in more recent literature but are mainly of historical interest. The remaining 4n + 1 series is not naturally occurring but comprises well-known artificial activities decaying down to stable thallium-205.

To extend the knowledge of nuclear binding energies, it is clearly necessary to make measurements to supplement the radioactive-decay energy cycles. In part, this extension can be made by measurement of Q values of artificial nuclear reactions. For example, the neutron-binding energies of the lead isotopes needed to link the energies of the four radioactive families together can be measured by determining the threshold gamma-ray energy to remove a neutron (photonuclear reaction); or the energies of incoming deuteron and outgoing proton in the reaction can be measured to provide this information.

Further extensions of nuclear-binding-energy measurements rely on precision mass spectroscopy. By ionizing, accelerating, and magnetically deflecting various nuclides, their masses can be measured with great precision. A precise measurement of the masses of atoms involved in radioactive decay is equivalent to direct measurement of the energy release in the decay process. The atomic mass of naturally occurring but radioactive potassium-40 is measured to be 39.964008 amu. Potassium-40 decays predominantly by β-emission to calcium-40, having a measured mass 39.962589. Through Einstein's equation, energy is equal to mass (m) times velocity of light (c) squared, or $E = mc^2$, the energy release (Q) and the mass difference, Δm, are related, the conversion factor being one amu, equal to 931.478 MeV. Thus, the excess mass of potassium-40 over calcium-40 appears as the total energy release Q_β in the radioactive decay Q_{β^-} = (39.964008 – 39.962589) × 931.478 MeV = 1.31 MeV. The other neighbouring isobar (same mass number, different atomic number) to argon-40 is also of lower mass, 39.962384, than potassium-40. This mass difference converted to energy units gives an energy release of 1.5 MeV, this being the energy release for EC decay to argon-40. The maximum energy release for positron emission is always less than that for electron capture by twice the rest mass energy of an electron ($2m_0c^2$ = 1.022 MeV); thus, the maximum positron energy for this reaction is 1.5 – 1.02, or 0.48 MeV.

To connect alpha-decay energies and nuclear mass differences requires a precise knowledge of the alpha-particle (helium-4) atomic mass. The mass of the parent minus the sum of the masses of the decay products gives the energy release. Thus, for alpha decay of plutonium-239 to uranium-235 and helium-4 the calculation goes as follows:

$$M(^{239}\text{Pu})\ \ 239.05216$$
$$-M(^{235}\text{U})\ \ 235.04393$$
$$-M(^{4}\text{He})\ -4.00260$$
$$\overline{\qquad 0.00563 \times 931.478 \qquad}$$
$$Q_\alpha = 5.24\,\text{MeV}$$

By combining radioactive-decay-energy information with nuclear-reaction Q values and precision mass spectroscopy, extensive tables of nuclear masses have been prepared. From them the Q values of unmeasured reactions or decay may be calculated.

Alternative to the full mass, the atomic masses may be expressed as mass defect, symbolized by the letter delta, Δ (the difference between the exact mass M and the integer A, the mass number), either in energy units or atomic mass units.

Absolute Nuclear Binding Energy

The absolute nuclear binding energy is the hypothetical energy release if a given nuclide were synthesized from Z separate hydrogen atoms and N (equal to A – Z) separate neutrons. An example is the calculation giving the absolute binding energy of the stablest of all nuclei, iron-56:

$$26 \times M(^1\text{H}) \qquad 26 \times 1.007825 = 26.20345$$
$$30 \times M(n) \qquad 30 \times 1.008665 = 30.25995$$
$$M(^{56}\text{Fe}) \qquad \qquad\qquad -55.93493$$
$$\overline{\qquad\qquad\qquad\qquad}$$
$$\text{Binding energy} = 0.52847 \times 931.478$$
$$= 492.58\,\text{MeV}$$

Average Binding energy:

Per nucleon of ^{56}Fe = 492.58/56 = 8.796 MeV

A general survey of the average binding energy per nucleon (for nuclei of all elements grouped according to ascending mass) shows a maximum at iron-56 falling off gradually on both sides to about 7 MeV at helium-4 and to about 7.4 MeV for the most massive nuclei known. Most of the naturally occurring nuclei are thus not stable in an absolute nuclear sense. Nuclei heavier than iron would gain energy by degrading into nuclear products closer to iron, but it is only for the elements of greatest mass that the rates of degradation processes such as alpha decay and spontaneous fission attain observable rates. In a similar manner, nuclear energy is to be gained by fusion of most elements lighter than iron. The coulombic repulsion between nuclei, however, keeps the rates of fusion reactions unobservably low unless the nuclei are subjected to temperatures of greater than 10^7 K. Only in the hot cores of the Sun and other stars or in thermonuclear bombs or controlled fusion plasmas are these temperatures attained and nuclear-fusion energy released.

Nuclear Models

The Liquid-drop Model

The average behaviour of the nuclear binding energy can be understood with the model of a charged liquid drop. In this model, the aggregate of nucleons has the same properties of a liquid drop, such as surface tension, cohesion, and deformation. There is a dominant attractive-binding-energy term proportional to the number of nucleons A. From this must be subtracted a surface-energy term proportional to surface area and a coulombic repulsion energy proportional to the square of the number of protons and inversely proportional to the nuclear radius. Furthermore, there is a symmetry-energy term of quantum-mechanical origin favouring equal numbers of protons and neutrons. Finally, there is a pairing term that gives slight extra binding to nuclei with even numbers of neutrons or protons.

The pairing-energy term accounts for the great rarity of odd–odd nuclei (the terms odd–odd, even–even, even–odd, and odd–even refer to the evenness or oddness of proton number, Z, and neutron number, N, respectively) that are stable against beta decay. The sole examples are deuterium, lithium-6, boron-10, and nitrogen-14. A few other odd–odd nuclei, such as potassium-40, occur in nature, but they are unstable with respect to beta decay. Furthermore, the pairing-energy term makes for the larger number of stable isotopes of even-Z elements, compared to odd-Z, and for the lack of stable isotopes altogether in element 43, technetium, and element 61, promethium.

The beta-decay energies of so-called mirror nuclei afford one means of estimating nuclear sizes. For example, the neon and fluorine nuclei, $^{19}_{10}Ne_9$ and $^{19}_9F_{10}$, are mirror nuclei because the proton and neutron numbers of one of them equal the respective neutron and proton numbers of the other. Thus, all binding-energy terms are the same in each except for the coulombic term, which is inversely proportional to the nuclear radius. Such calculations along with more direct determinations by high-energy electron scattering and energy measurements of X-rays from muonic atoms (hydrogen atoms in which the electrons are replaced by negative muons) establish the nuclear charge as roughly uniformly distributed in a sphere of radius 1.2 $A^{1/3} \times 10^{-13}$ centimetre. That the radius is proportional to the cube root of the mass number has the great significance that the average density of all nuclei is nearly constant.

Careful examination of nuclear-binding energies reveals periodic deviations from the smooth average behaviour of the charged-liquid-drop model. An extra binding energy arises in the neighbourhood of certain numbers of neutrons or protons, the so-called magic numbers (2, 8, 20, 28, 50, 82, and 126). Nuclei such as $^4_2He_2$, $^{16}_8O_8$, $^{40}_{20}Ca_{20}$, $^{48}_{20}Ca_{28}$ and $^{208}_{82}Pb_{126}$ are especially stable species, doubly magic, in view of their having both proton and neutron numbers magic. These doubly magic nuclei are situated at the intersections of grid lines on figure above.

The Shell Model

There were noted periodic binding-energy irregularities at the magic numbers. The periodic occurrence of magic numbers of extra stability is strongly analogous to the extra electronic stabilities occurring at the atomic numbers of the noble-gas atoms. The explanations of these stabilities are quite analogous in atomic and nuclear cases as arising from filling of particles into quantized orbitals of motion. The completion of filling of a shell of orbitals is accompanied by an extra stability. The nuclear model accounting for the magic numbers is, as previously noted, the shell model. In its simplest form, this model can account for the occurrence of spin zero for all even–even nuclear ground states; the nucleons fill pairwise into orbitals with angular momenta canceling. The shell model also readily accounts for the observed nuclear spins of the odd-mass nuclei adjacent to doubly magic nuclei, such as $^{207}_{82}Pb$. Here, the spins of 1/2 for neighbouring $^{207}_{81}Tl$ and $^{207}_{82}Pb$ are accounted for by having all nucleons fill pairwise into the lowest energy orbits and putting the odd nucleon into the last available orbital before reaching the doubly magic configuration (the Pauli exclusion principle dictates that no more than two nucleons may occupy a given orbital, and their spins must be oppositely directed); calculations show the last available orbitals below lead-208 to have angular momentum 1/2. Likewise, the spins of 9/2 for $^{207}_{82}Pb$ and $^{207}_{83}Bi$ are understandable because spin-9/2 orbitals are the next available orbitals beyond doubly magic lead-208. Even the associated magnetization, as expressed by the magnetic dipole moment, is rather well explained by the simple spherical-shell model.

The orbitals of the spherical-shell model are labeled in a notation close to that for electronic orbitals in atoms. The orbital configuration of calcium-40 has protons and neutrons filling the following orbitals: $1s_{1/2}$, $1p_{3/2}$, $1p_{1/2}$, $1d_{5/2}$, and $1d_{3/2}$. The letter denotes the orbital angular momentum in usual spectroscopic notation, in which the letters s, p, d, f, g, h, i, etc., represent integer values of l running from zero for s through six for i. The fractional subscript gives the total angular momentum j with values of $l + 1/2$ and $l - 1/2$ allowed, as the intrinsic spin of a nucleon is 1/2. The first integer is a radial quantum number taking successive values 1, 2, 3, etc., for successively higher energy values of an orbital of given l and j. Each orbital can accommodate a maximum of $2j + 1$ nucleons. The exact order of various orbitals within a shell differs somewhat for neutrons and protons. The parity associated with an orbital is even (+) if l is even (s, d, g, i) and odd (−) if l is odd (p, f, h).

Table: Spherical-shell-model Orbitals.

Shell Closure Number	
2	$1s_{1/2}$
8	$1p_{3/2}$, $1p$
20	$1d_{5/2}$, $2s_{1/2}$, $1d_{3/2}$
28	$1f_{7/2}$
50	$2p_{3/2}$, $1f_{5/2}$, $2p_{1/2}$, $1g_{9/2}$
82	$1g_{7/2}$, $2d_{5/2}$, $1h_{11/2}$, $2d_{3/2}$, $3s_{1/2}$
126	$2f_{7/2}$, $1h_{9/2}$, $1i_{13/2}$, $3p_{3/2}$, $2f_{5/2}$, $3p_{1/2}$
184	$2g_{9/2}$, $1i_{11/2}$, $1j_{15/2}$, $3d_{5/2}$, $2g_{7/2}$, $4s_{1/2}$, $3d_{3/2}$

An example of a spherical-shell-model interpretation is provided by the beta-decay scheme of 2.2-minute thallium-209 shown, in which spin and parity are given for each state. The ground and lowest excited states of lead-209 are to be associated with occupation by the 127th neutron of the lowest available orbitals above the closed shell of 126. From the last line of the table, it is to be noted that there are available $g_{9/2}$, $d_{5/2}$, and $s_{1/2}$ orbitals with which to explain the ground and first two excited states. Low-lying states associated with the $i_{11/2}$ and $j_{15/2}$ orbitals are known from nuclear-reaction studies, but they are not populated in the beta decay.

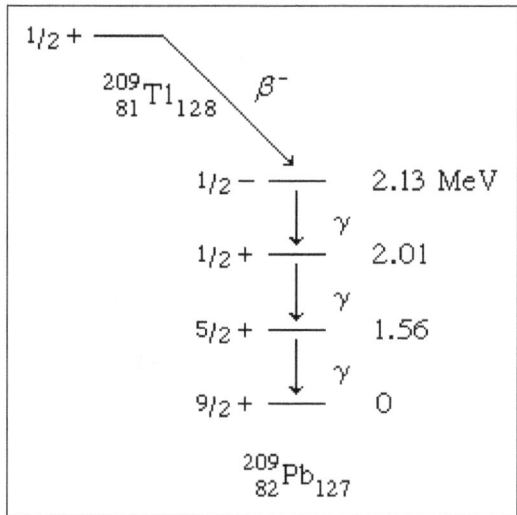

The 2.13-MeV state that receives the primary beta decay is not so simply interpreted as the other states. It is to be associated with the promotion of a neutron from the $3p_{1/2}$ orbital below the 126 shell closure. The density (number per MeV) of states increases rapidly above this excitation, and the interpretations become more complex and less certain.

By suitable refinements, the spherical-shell model can be extended further from the doubly magic region. Primarily, it is necessary to drop the approximation that nucleons move independently in orbitals and to invoke a residual force, mainly short-range and attractive, between the nucleons. The spherical-shell model augmented by residual interactions can explain and correlate around the magic regions a large amount of data on binding energies, spins, magnetic moments, and the spectra of excited states.

Collective Model

For nuclei more removed from the doubly magic regions, the spherical-shell model encounters difficulty in explaining the large observed electric quadrupole moments indicating cigar-shaped nuclei. For these nuclei a hybrid of liquid-drop and shell models, the collective model, has been proposed.

Nucleons can interact with one another in a collective fashion to deform the nuclear shape to a cigar shape. Such large spheroidal distortions are usual for nuclei far from magic, notably with $150 \lesssim A \lesssim 190$, and $224 \lesssim A$ (the symbol < denotes less than, and ~ means that the number is approximate). In these deformed regions the collective model prescribes that orbitals be computed in a cigar-shaped potential and that the relatively low-energy rotational excitations of the tumbling motion of the cigar shape be taken into account. The collective model has been highly successful in correlating and predicting nuclear properties in deformed regions. An example of a nuclear rotational band (a series of adjacent states) is provided by the decay of the isomer hafnium-180m, in figure, through a cascade of gamma rays down the ground rotational band.

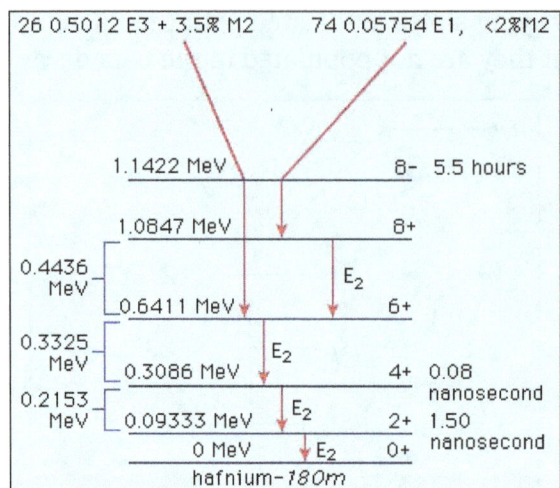

The decay scheme of hafnium-180*m*.

Rates of Radioactive Transitions

There is a vast range of the rates of radioactive decay, from undetectably slow to unmeasurably short. Before considering the factors governing particular decay rates in detail, it seems appropriate to review the mathematical equations governing radioactive decay and the general methods of rate measurement in different ranges of half-life.

Exponential-decay Law

Radioactive decay occurs as a statistical exponential rate process. That is to say, the number of atoms likely to decay in a given infinitesimal time interval (dN/dt) is proportional to the number (*N*) of atoms present. The proportionality constant, symbolized by the Greek letter lambda, λ, is called the decay constant. Mathematically, this statement is expressed by the first-order differential equation,

$$-\frac{dN}{dt} = \lambda N$$

This equation is readily integrated to give,

$$N(t) = N_0 e^{-\lambda t}$$

in which N_0 is the number of atoms present when time equals zero. From the above two equations it may be seen that a disintegration rate, as well as the number of parent nuclei, falls exponentially with time. An equivalent expression in terms of half-life $t_{1/2}$ is,

$$N(t) = N_0 (1/2)^r$$
$$r = t / t_{1/2}$$

It can readily be shown that the decay constant λ and half-life ($t_{1/2}$) are related as follows: $\lambda = \log_e 2 / t_{1/2} = 0.693/t_{1/2}$. The reciprocal of the decay constant λ is the mean life, symbolized by the letter tau, τ.

For a radioactive nucleus such as potassium-40 that decays by more than one process (89 percent β–, 11 percent electron capture), the total decay constant is the sum of partial decay constants for each decay mode. (The partial half-life for a particular mode is the reciprocal of the partial decay constant times 0.693.) It is helpful to consider a radioactive chain in which the parent (generation 1) of decay constant λ_1 decays into a radioactive daughter (generation 2) of decay constant λ_2. The case in which none of the daughter isotope (2) is originally present yields an initial growth of daughter nuclei followed by its decay. The equation giving the number (N_2) of daughter nuclei existing at time t in terms of the number $N_1(0)$ of parent nuclei present when time equals zero is:

$$N_2(t) = \lambda_1 N_1(0) \frac{e^{-\lambda 1 t} - e^{-\lambda 2 t}}{\lambda_2 - \lambda_1}$$

in which e represents the logarithmic constant 2.71828.

The general equation for a chain of n generations with only the parent initially present (when time equals zero) is as follows:

$$N_n(t) = N_1(0)(C_1 e^{-\lambda_1 t} + C_1 e^{-\lambda_2 t} + \ldots\ldots\ldots + C_n e^{-\lambda_n t}) \lambda_1 \lambda_2 \ldots\ldots\ldots \lambda_{n-1}$$

in which e represents the logarithmic constant 2.71828.

$$C_1 = \frac{1}{(\lambda_2 - \lambda_1)(\lambda_3 - \lambda_1)\ldots\ldots\ldots(\lambda_n - \lambda_1)}$$

$$C_2 = \frac{1}{(\lambda_1 - \lambda_2)(\lambda_3 - \lambda_2)\ldots\ldots\ldots(\lambda_n - \lambda_2)}$$

$$C_n = \frac{1}{(\lambda_1 - \lambda_n)(\lambda_2 - \lambda_n)\ldots\ldots\ldots(\lambda_{n-1} - \lambda_n)}$$

These equations can readily be modified to the case of production of isotopes in the steady neutron flux of a reactor or in a star. In such cases, the chain of transformations might be mixed with some steps occurring by neutron capture and some by radioactive decay. The neutron-capture probability for a nucleus is expressed in terms of an effective cross-sectional area. If one imagines the nuclei replaced by spheres of the same cross-sectional area, the rate of reaction in a neutron flux would be given by the rate at which neutrons strike the spheres. The cross section is usually symbolized by the Greek letter sigma, σ, with the units of barns (10^{-24} cm^2) or millibarns (10^{-3} b) or microbarns (10^{-6} b). Neutron flux is often symbolized by the letters nv (neutron density, n, or number per cubic centimetre, times average speed, v) and given in neutrons per square centimetre per second.

The modification of the transformation equations merely involves substituting the product $nv\sigma_i$ in place of λ_i for any step involving neutron capture rather than radioactive decay. Reactor fluxes nv even higher than 10^{15} neutrons per square centimetre per second are available in several research reactors, but usual fluxes are somewhat lower by a factor of 1,000 or so. Tables of neutron-capture cross sections of the naturally occurring nuclei and some radioactive nuclei can be used for calculation of isotope production rates in reactors.

Measurement of Half-life

The measurement of half-lives of radioactivity in the range of seconds to a few years commonly involves measuring the intensity of radiation at successive times over a time range comparable to the half-life. The logarithm of the decay rate is plotted against time, and a straight line is fitted to the points. The time interval for this straight-line decay curve to fall by a factor of 2 is read from the graph as the half-life, by virtue of equations. If there is more than one activity present in the sample, the decay curve will not be a straight line over its entire length, but it should be resolvable graphically (or by more sophisticated statistical analysis) into sums and differences of straight-line exponential terms. The general equations for chain decays show a time dependence given by sums and differences of exponential terms, though special modified equations are required in the unlikely case that two or more decay constants are identically equal.

$$-\frac{dN}{dt} = \lambda N$$

$$N_t = N_0 e^{-\lambda t}$$

For half-lives longer than several years it is often not feasible to measure accurately the decrease in counting rate over a reasonable length of time. In such cases, a measurement of specific activity may be resorted to; i.e., a carefully weighed amount of the radioactive isotope is taken for counting measurements to determine the disintegration rate, D. Then by equation above the decay constant λi may be calculated. Alternately, it may be possible to produce the activity of interest in such a way that the number of nuclei, N, is known, and again with a measurement of D equation above may be used. The number of nuclei, N, might be known from counting the decay of a parent activity or from knowledge of the production rate by a nuclear reaction in a reactor or accelerator beam.

Half-lives from 100 microseconds to one nanosecond are measured electronically in coincidence experiments. The radiation yielding the species of interest is detected to provide a start pulse for an electronic clock, and the radiation by which the species decays is detected in another device to provide a stop pulse. The distribution of these time intervals is plotted semi-logarithmically, as discussed for the decay-rate treatment, and the half-life is determined from the slope of the straight line.

Half-lives in the range of 100 microseconds to one second must often be determined by special techniques. For example, the activities produced may be deposited on rapidly rotating drums or moving tapes, with detectors positioned along the travel path. The activity may be produced so as to travel through a vacuum at a known velocity and the disintegration rate measured as a function of distance; however, this method usually applies to shorter half-lives in or beyond the range of the electronic circuit.

Species with half-lives shorter than the electronic measurement limit are not considered as separate radioactivities, and the various techniques of determining their half-lives will hence not be cited here.

Alpha Decay

Alpha decay, the emission of helium ions, exhibits sharp line spectra when spectroscopic measurements of the alpha-particle energies are made. For even–even alpha emitters the most intense alpha group or line is always that leading to the ground state of the daughter. Weaker lines of lower energy go to excited states, and there are frequently numerous lines observable.

The main decay group of even–even alpha emitters exhibits a highly regular dependence on the atomic number, Z, and the energy release, Q_α. (Total alpha energy release, Q_α, is equal to alpha-particle energy, E_α, plus daughter recoil energy needed for conservation of momentum; $E_{recoil} = (m_\alpha/[m_\alpha + M_d])E_\alpha$, with m_α equal to the mass of the alpha particle and Md the mass of the daughter product.) As early as 1911 the German physicist Johannes Wilhelm Geiger, together with the British physicist John Mitchell Nuttall, noted the regularities of rates for even–even nuclei and proposed a remarkably successful equation for the decay constant, $\log \lambda = a + b \log r$, in which r is the range in air, b is a constant, and a is given different values for the different radioactive series. The decay constants of odd alpha emitters (odd A or odd Z or both) are not quite so regular and may be much smaller. The values of the constant b that were used by Geiger and Nuttall implied a roughly 90th-power dependence of λ on Q_α. There is a tremendous range of known half-lives from the 2×10^{15} years of $^{144}_{60}Nd$ (neodymium) with its 1.83-MeV alpha-particle energy (E_α) to the 0.3 microsecond of $^{212}_{84}Po$ (polonium) with $E_\alpha = 8.78$ MeV.

The theoretical basis for the Geiger–Nuttall empirical rate law remained unknown until the formulation of wave mechanics. A dramatic early success of wave mechanics was the quantitative theory of alpha-decay rates. One curious feature of wave mechanics is that particles may have a nonvanishing probability of being in regions of negative kinetic energy. In classical mechanics a ball that is tossed to roll up a hill will slow down until its gravitational potential energy equals its total energy, and then it will roll back toward its starting point. In quantum mechanics the ball has a certain probability of tunneling through the hill and popping out on the other side. For objects large enough to be visible to the eye, the probability of tunneling through energetically forbidden regions is unobservably small. For submicroscopic objects such as alpha particles, nucleons, or electrons, however, quantum mechanical tunneling can be an important process—as in alpha decay.

The logarithm of tunneling probability on a single collision with an energy barrier of height B and thickness D is a negative number proportional to thickness D, to the square root of the product of B and particle mass m. The size of the proportionality constant will depend on the shape of the barrier and will depend inversely on Planck's constant h.

In the case of alpha decay, the electrostatic repulsive potential between alpha particle and nucleus generates an energetically forbidden region, or potential barrier, from the nuclear radius out to several times this distance. The maximum height (B) of this alpha barrier is given approximately by the expression $B = 2Ze^2/R$, in which Z is the charge of the daughter nucleus, e is the elementary charge in electrostatic units, and R is the nuclear radius. Numerically, B is roughly equal to $2Z/A^{1/3}$, with A the mass number and B in energy units of MeV. Thus, although the height of the potential barrier for $^{212}_{84}Po$ decay is nearly 28 MeV, the total energy released is $Q_\alpha = 8.95$ MeV. The thickness of the barrier (i.e., distance of the alpha particle from the centre of the nucleus at the moment of recoil) is about twice the nuclear radius of 8.8×10^{-13} centimetre. The tunneling calculation for the transition probability (P) through the barrier gives approximately:

$$P = \exp\left[\left(-\frac{\sqrt{2MBR}}{\hbar}\right)\left(\frac{\pi B^{1/2}}{Q^{1/2}} - 4\right)\right]$$

in which M is the mass of the alpha particle and $\hbar$ is Planck's constant h divided by 2π. By making simple assumptions about the frequency of the alpha particle striking the barrier, the penetration formula can be used to calculate an effective nuclear radius for alpha decay. This method was the early ways of estimating nuclear sizes. In more sophisticated modern techniques the radius value is taken from other experiments, and alpha-decay data and penetrabilities are used to calculate the frequency factor.

The form of equation above suggests the correlation of decay rates by an empirical expression relating the half-life $(t_{1/2})$ of decay in seconds to the release energy (Q_α) in MeV:

$$\log t_{1/2} = \frac{a}{\sqrt{Q_a}} + b$$

Values of the constants a and b that give best fits to experimental rates of even–even nuclei with neutron number greater than 126 are given in the table. The nuclei with 126 or fewer neutrons decay more slowly than the heavier nuclei, and constants a and b must be readjusted to fit their decay rates.

Table: Semiempirical constants.

	a	b
98 californium (Cf)	152.86	−52.9506
96 curium (Cm)	152.44	−53.6825
94 plutonium (Pu)	146.23	−52.0899
92 uranium (U)	147.49	−53.6565
90 thorium (Th)	144.19	−53.2644
88 radium (Ra)	139.17	−52.1476
86 radon (Rn)	137.46	−52.4597
84 polonium (Po)	129.35	−49.9229

The alpha-decay rates to excited states of even–even nuclei and to ground and excited states of nuclei with odd numbers of neutrons, protons, or both may exhibit retardations from equation above rates ranging to factors of thousands or more. The factor by which the rate is slower than the rate formula above is the hindrance factor. The existence of uranium-235 in nature rests on the fact that alpha decay to the ground and low excited states exhibits hindrance factors of over 1,000. Thus the uranium-235 half-life is lengthened to 7×10^8 years, a time barely long enough compared to the age of the elements in the solar system for uranium-235 to exist in nature today.

The alpha hindrance factors are fairly well understood in terms of the orbital motion of the individual protons and neutrons that make up the emitted alpha particle. The alpha-emitting nuclei heavier than radium are considered to be cigar-shaped, and alpha hindrance factor data have been used to infer the most probable zones of emission on the nuclear surface—whether polar, equatorial, or intermediate latitudes.

Beta Decay

The processes separately introduced at the beginning of this section as beta-minus decay, beta-plus decay, and orbital electron capture can be appropriately treated together. They all are processes whereby neutrons and protons may transform to one another by weak interaction. In striking contrast to alpha decay, the electrons (minus or plus charged) emitted in beta-minus and beta-plus decay do not exhibit sharp, discrete energy spectra but have distributions of electron energies ranging from zero up to the maximum energy release, Q_β. Furthermore, measurements of heat released by beta emitters (most radiation stopped in surrounding material is converted into heat energy) show a substantial fraction of the energy, Q_β is missing. These observations, along with other considerations involving the spins or angular momenta of nuclei and electrons, led Wolfgang Pauli to postulate the simultaneous emission of the neutrino. The neutrino, as a light and uncharged particle with nearly no interaction with matter, was supposed to carry off the missing heat energy. Today, neutrino theory is well accepted with the elaboration that there are six kinds of neutrinos, the electron neutrino, mu neutrino, and tau neutrino and corresponding antineutrinos of each. The electron neutrinos are involved in nuclear beta-decay transformations, the mu neutrinos are encountered in decay of muons to electrons, and the tau neutrinos are produced when a massive lepton called a tau breaks down.

Although in general the more energetic the beta decay the shorter is its half-life, the rate relationships do not show the clear regularities of the alpha-decay dependence on energy and atomic number.

The first quantitative rate theory of beta decay was given by Enrico Fermi in 1934, and the essentials of this theory form the basis of modern theory. As an example, in the simplest beta-decay process, a free neutron decays into a proton, a negative electron, and an antineutrino: $n \rightarrow p + e^- \; \bar{\nu}$ the weak interaction responsible for this process, in which there is a change of species (n to p) by a nucleon with creation of electron and antineutrino, is characterized in Fermi theory by a universal constant, g. The sharing of energy between electron and antineutrino is governed by statistical probability laws giving a probability factor for each particle proportional to the square of its linear momentum (defined by mass times velocity for speeds much less than the speed of light and by a more complicated, relativistic relation for faster speeds). The overall probability law from Fermi theory gives the probability per unit time per unit electron energy interval, $P(W)$, as follows:

$$P(W) = \frac{64\pi^4 m_0^5 c^4 g^2}{h^7} W (W^2 - 1)^{1/2} (W_0 - W)^2$$

in which W is the electron energy in relativistic units ($W = 1 + E/m_{oc}^2$) and W_0 is the maximum ($W_0 = 1 + Q_\beta/m_{oc}^2$), m_0 the rest mass of the electron, c the speed of light, and h Planck's constant. This rate law expresses the neutron beta-decay spectrum in good agreement with experiment, the spectrum falling to zero at lowest energies by the factor W and falling to zero at the maximum energy by virtue of the factor $(W_0 - W)^2$.

In Fermi's original formulation, the spins of an emitted beta and neutrino are opposing and so cancel to zero. Later work showed that neutron beta decay partly proceeds with the $1/2$ $\hbar$ spins of beta and neutrino adding to one unit of $\hbar$. The former process is known as Fermi decay (F) and the latter Gamow–Teller (GT) decay, after George Gamow and Edward Teller, the physicists who first proposed it. The interaction constants are determined to be in the ratio $g_{GT}^2/g_F^2 = 1.4$. Thus, g^2 in equation should be replaced by $(g_F^2 + g_{GT}^2)$.

The scientific world was shaken in 1957 by the measurement in beta decay of maximum violation of the law of conservation of parity. The meaning of this nonconservation in the case of neutron beta decay considered above is that the preferred direction of electron emission is opposite to the direction of the neutron spin. By means of a magnetic field and low temperature it is possible to cause neutrons in cobalt-60 and other nuclei, or free neutrons, to have their spins set preferentially in the up direction perpendicular to the plane of the coil generating the magnetic field. The fact that beta decay prefers the down direction for spin means that the reflection of the experiment as seen in a mirror parallel to the coil represents an unphysical situation: conservation of parity, obeyed by most physical processes, demands that experiments with positions reversed by mirror reflection should also occur. Further consequences of parity violation in beta decay are that spins of emitted neutrinos and electrons are directed along the direction of flight, totally so for neutrinos and partially so by the ratio of electron speed to the speed of light for electrons.

The overall half-life for beta decay of the free neutron, measured as 12 minutes, may be related to the interaction constants g^2 (equal to $g_F^2 + g_{GT}^2$) by integrating (summing) probability expression over all possible electron energies from zero to the maximum. The result for the decay constant is:

$$\lambda = \frac{64\pi^4 m_0^5 c^4 g^2}{h^7} \left\{ (W_0^2 - 1)^{1/2} \left(\frac{W_0^4}{30} - \frac{3W_0^2}{20} - \frac{2}{15} \right) + \frac{W_0}{4} \ln\left[W_0 \left(W_0^2 - 1 \right)^{1/2} \right] \right\}$$

in which W_0 is the maximum beta-particle energy in relativistic units ($W_0 = 1 + Q_\beta/m_{oc}^2$), with m_0 the rest mass of the electron, c the speed of light, and h Planck's constant. The best g value from decay rates is approximately 10^{-49} erg per cubic centimetre. As may be noted from equation above, there is a limiting fifth-power energy dependence for highest decay energies.

In the case of a decaying neutron not free but bound within a nucleus, the above formulas must be modified. First, as the nuclear charge Z increases, the relative probability of low-energy electron emission increases by virtue of the coulombic attraction. For positron emission, which is energetically impossible for free protons but can occur for bound protons in proton-rich nuclei, the nuclear coulomb charge suppresses lower energy positrons from the shape given by equation given above. This equation can be corrected by a factor $F(Z,W)$ depending on the daughter atomic number Z and electron energy W. The factor can be calculated quantum mechanically. The coulomb charge also affects the overall rate expression (above) such that it can no longer be expressed as an algebraic function, but tables are available for analysis of beta decay rates. The rates are analyzed in terms of a function $f(Z,Q_\beta)$ calculated by integration of equation (above) with correction factor $F(Z,W)$.

Approximate expressions for the f functions usable for decay energies Q between 0.1 MeV and 10 MeV, in which Q is measured in MeV, and Z is the atomic number of the daughter nucleus, are as follows (the symbol $\approx$ means approximately equal to):

$$f_{\beta-} \approx 6.0Q^{4-0.005(Z-1)} \cdot 10^{Z/50}$$

$$f_{\beta+} \approx 6.2Q^4 10^{0.007Z} \cdot 10^{0.009Z(\log 1/3Q)^2}$$

For electron capture, a much weaker dependence on energy is found:

$$f_{EC} \approx (Z+1)^{3.5} Q / 4 \times 10^5$$

The basic beta decay rate expression obeyed by the class of so-called superallowed transitions, including decay of the neutron and several light nuclei is:

$$\lambda_\beta = \frac{64\pi m_0^5 c^4 g^2}{h^7}$$

Like the ground-to-ground alpha transitions of even–even nuclei, the superallowed beta transitions obey the basic rate law, but most beta transitions go much more slowly. The extra retardation is explained in terms of mismatched orbitals of neutrons and protons involved in the transition. For the superallowed transitions the orbitals in initial and final states are almost the same. Most of them occur between mirror nuclei, with one more or less neutron than protons; i.e., beta-minus decay of hydrogen-3, electron capture of beryllium-7 and positron emission of carbon-11, oxygen-15, neon-19 and titanium-43.

The nuclear retardation of beta decay rates below those of the superallowed class may be expressed in a fundamental way by multiplying the right side of equation below by the square of a nuclear matrix element (a quantity of quantum mechanics), which may range from unity down to zero depending on the degree of mismatch of initial and final nuclear states of internal motion. A more usual way of expressing the nuclear factor of the beta rate is the log ft value, in which f refers to the function $f(Z,Q_\beta)$. Because the half-life is inversely proportional to the decay constant λ, the product $f_{\beta t1/2}$ will be a measure of (inversely proportional to) the square of the nuclear matrix element. For the log ft value, the beta half-life is taken in seconds, and the ordinary logarithm to the base 10 is used. The superallowed transitions have log ft values in the range of 3 to 3.5. Beta log ft values are known up to as large as ~ 23 in the case of indium-115. There is some correlation of log ft values with spin changes between parent and daughter nucleons, the indium-115 decay involving a spin change of four, whereas the superallowed transitions all have spin changes of zero or one.

$$\lambda_\beta = \frac{64\pi m_0^5 c^4 g^2}{h^7} f_\beta$$

Gamma Transition

The nuclear gamma transitions belong to the large class of electromagnetic transitions encompassing radio-frequency emission by antennas or rotating molecules, infrared emission by vibrating molecules or hot filaments, visible light, ultraviolet light, and X-ray emission by electronic jumps in atoms or molecules. The usual relations apply for connecting frequency v, wavelength λ, and photon quantum energy E with speed of light c and Planck's constant h; namely, $\lambda = c/v$ and $E = hv$. It is sometimes necessary to consider the momentum (p) of the photon given by $p = E/c$.

Classically, radiation accompanies any acceleration of electric charge. Quantum mechanically there is a probability of photon emission from higher to lower energy nuclear states, in which the internal state of motion involves acceleration of charge in the transition. Therefore, purely neutron orbital acceleration would carry no radiative contribution.

A great simplification in nuclear gamma transition rate theory is brought about by the circumstance that the nuclear diameters are always much smaller than the shortest wavelengths of gamma radiation in radioactivity—i.e., the nucleus is too small to be a good antenna for the radiation. The simplification is that nuclear gamma transitions can be classified according to multipolarity, or amount of spin angular momentum carried off by the radiation. One unit of angular momentum in the radiation is associated with dipole transitions (a dipole consists of two separated equal charges, plus and minus). If there is a change of nuclear parity, the transition is designated electric dipole (E1) and is analogous to the radiation of a linear half-wave dipole radio antenna. If there is no parity change, the transition is magnetic dipole (M1) and is analogous to the radiation of a full-wave loop antenna. With two units of angular momentum change, the transition is electric quadrupole (E2), analogous to a full-wave linear antenna of two dipoles out-of-phase, and magnetic quadrupole (M2), analogous to coaxial loop antennas driven out-of-phase. Higher multipolarity radiation also frequently occurs with radioactivity.

Transition rates are usually compared to the single-proton theoretical rate, or Weisskopf formula, named after the American physicist Victor Frederick Weisskopf, who developed it. The table gives the theoretical reference rate formulas in their dependence on nuclear mass number A and gamma-ray energy $E\gamma$ (in MeV).

Table: Gamma transition rates.

Transition Type	Partial Half-life t_γ (seconds)	Illustrative t_γ values for A = 125, E = 0.1 MeV (seconds)
E1	$5.7 \times 10^{-15} \, E^{-3} \, A^{-2/3}$	2×10^{-13}
E2	$6.7 \times 10^{-9} \, E^{-5} \, A^{-4/3}$	1×10^{-6}
E3	$1.2 \times 10^{-2} \, E^{-7} \, A^{-2}$	8
E4	$3.4 \times 10^{4} \, E^{-9} \, A^{-8/3}$	9×10^{7}
E5	$1.3 \times 10^{11} \, E^{-11} \, A^{-10/3}$	1×10^{15}
M1	$2.2 \times 10^{-14} \, E^{-3}$	2×10^{-11}
M2	$2.6 \times 10^{-8} \, E^{-5} \, A^{-2/3}$	1×10^{-4}
M3	$4.9 \times 10^{-2} \, E^{-7} \, A^{-4/3}$	8×10^{2}
M4	$1.3 \times 10^{5} \, E^{-9} \, A^{-2}$	8×10^{9}
M5	$5.0 \times 10^{11} \, E^{-11} \, A^{-8/3}$	1×10^{17}

It is seen for the illustrative case of gamma energy 0.1 MeV and mass number 125 that there occurs an additional factor of 10^7 retardation with each higher multipole order. For a given multipole, magnetic radiation should be a factor of 100 or so slower than electric. These rate factors ensure that nuclear gamma transitions are nearly purely one multipole, the lowest permitted by the nuclear spin change. There are many exceptions, however; mixed M1–E2 transitions are common, because E2 transitions are often much faster than the Weisskopf formula gives and M1 transitions are generally slower. All E1 transitions encountered in radioactivity are much slower than the Weisskopf formula. The other higher multipolarities show some scatter in rates, ranging from agreement to considerable retardation. In most cases the retardations are well understood in terms of nuclear model calculations.

Though not literally a gamma transition, electric monopole (E0) transitions may appropriately be mentioned. These may occur when there is no angular momentum change between initial and final nuclear states and no parity change. For spin-zero to spin-zero transitions, single gamma emission is strictly forbidden. The electric monopole transition occurs largely by the ejection of electrons from the orbital cloud in heavier elements and by positron–electron pair creation in the lighter elements.

Nuclear Physics

Nuclear physics is a field of physics that involves investigation of the building blocks and interactions of atomic nuclei. It includes studies of nuclear components such as protons and neutrons, forces such as the strong force (or strong interaction), and phenomena such as radioactive decay, nuclear fission, and nuclear fusion.

Nuclear power and nuclear weapons are the most commonly known applications of nuclear physics, but the research field is also the basis for a far wider range of less common applications, such as in medicine (nuclear medicine, magnetic resonance imaging), materials engineering (ion implantation), and archaeology (radiocarbon dating).

Related Fields

Nuclear physics is sometimes used synonymously with atomic physics, but physicists usually differentiate between the two. Atomic physics studies the combined system of the atomic nucleus and the arrangement of electrons around the nucleus.

Particle physics involves study of the elementary constituents of matter and radiation, and the interactions between them. Particle physics evolved out of nuclear physics and, for this reason, has been included under the same term in earlier times.

Rutherford's Team Discovers the Nucleus

In 1906, Ernest Rutherford published "Radiation of the α Particle from Radium in passing through Matter." Hans Geiger expanded on this work in a communication to the Royal Society with experiments he and Rutherford had done passing α particles through air, aluminum foil and gold leaf. More work was published in 1909, by Geiger and Ernest Marsden, and was greatly expanded in 1910 by Geiger. In 1911-12, Rutherford went before the Royal Society to explain the experiments and propound the new theory of the atomic nucleus as we now understand it.

The key experiment behind this announcement was performed in 1909, when Hans Geiger and Ernest Marsden, under Rutherford's supervision, fired alpha particles (helium nuclei) at a thin film of gold foil. The plum pudding model predicted that the alpha particles should come out of the foil with their trajectories being at most slightly bent. Their actual observations, however, were shocking: A few particles were scattered through large angles, with some bouncing completely backwards.

Those observations, upon analysis, led to the Rutherford model of the atom, in which the atom has a very small, dense nucleus containing most of its mass, and consisting of heavy positively charged particles with several embedded electrons that would (at least partially) balance out the charge (since the neutron was unknown). As an example of this model (which is not the modern one), nitrogen-14 was thought to consist of a nucleus with 14 protons and 7 electrons (21 total particles), and the nucleus was surrounded by 7 more orbiting electrons.

The Rutherford model worked quite well until studies of nuclear spin was carried out by Franco Rasetti at the California Institute of Technology in 1929. By 1925, it was known that protons and electrons had a spin of 1/2, and in the Rutherford model of nitrogen-14, 20 of the 21 particles should have paired up to cancel each other's spin, and the final odd particle should have left the nucleus with a spin of 1/2. Rasetti discovered, however, that nitrogen-14 has a spin of 1.

Chadwick Discovers the Final Necessary Particle

In 1932, Chadwick realized that radiation that had been observed by Walther Bothe, Herbert L. Becker, and Irène and Frédéric Joliot-Curie was actually due to a neutral particle of about the same mass as the proton, that he called the neutron (following a suggestion about the need for such a particle, by Rutherford). That same year, Dmitri Ivanenko suggested that neutrons were in fact spin 1/2 particles and that the nucleus contained neutrons to explain the mass not due to protons, and that there were no electrons in the nucleus—only protons and neutrons. The neutron spin immediately solved the problem of the spin of nitrogen-14, as the one unpaired proton and one unpaired neutron in this model, each contribute a spin of 1/2 in the same direction, for a final total spin of 1.

With the discovery of the neutron, scientists could at last calculate what fraction of binding energy each nucleus had, from comparing the nuclear mass with that of the protons and neutrons that composed it. Differences between nuclear masses calculated in this way, and when nuclear reactions were measured, where found to agree with Einstein's calculation of the equivalence of mass and energy to high accuracy.

Yukawa's Meson Postulated to Bind Nuclei

In 1935, Hideki Yukawa proposed the first significant theory of the strong force to explain how neutrons and protons are held together in the nucleus. In the Yukawa interaction, a virtual particle, later called a meson, mediated a force between all nucleons, including protons and neutrons. This force explained why nuclei did not disintegrate under the influence of proton repulsion, and it also gave an explanation of why the attractive strong force had a more limited range than the electromagnetic repulsion between protons. Later, the discovery of the pi meson showed it to have the properties of Yukawa's particle.

With Yukawa's papers, the modern model of the atom was nearing completion. The center of the atom contains a tight ball of neutrons and protons, which is held together by the strong nuclear force, unless the nucleus is too large. Unstable nuclei may undergo alpha decay, when they emit an energetic helium nucleus, or beta decay, when they eject an electron (or positron). After one of these decays, the resultant nucleus may be left in an excited state, and in this case it decays to its ground state by emitting high energy photons (gamma decay).

The study of the strong and weak nuclear forces (the latter explained by Enrico Fermi via Fermi's interaction in 1934) led physicists to collide nuclei and electrons at ever higher energies. This research became the science of particle physics, the crown jewel of which is the standard model of particle physics, which unifies the strong, weak, and electromagnetic forces.

Modern Nuclear Physics

A heavy nucleus can contain hundreds of nucleons, which means that with some approximation it can be treated as a classical system, rather than a quantum-mechanical one. In the resulting liquid-drop model, the nucleus has an energy that arises partly from surface tension and partly from electrical repulsion of the protons. The liquid-drop model is able to reproduce many features of nuclei, including the general trend of binding energy with respect to mass number, as well as the phenomenon of nuclear fission.

Superimposed on this classical picture, however, are quantum mechanical effects, which can be described using the nuclear shell model, developed in large part by Maria Goeppert-Mayer. Nuclei with certain numbers of neutrons and protons (the magic numbers 2, 8, 20, 50, 82, 126, ...) are particularly stable, because their shells are filled.

Other, more complicated models for the nucleus have also been proposed, such as the interacting boson model, in which pairs of neutrons and protons interact as bosons, analogously to Cooper pairs of electrons.

Much of current research in nuclear physics relates to the study of nuclei under extreme conditions, such as high spin and excitation energy. Nuclei may also have extreme shapes (similar to that of American footballs) or extreme neutron-to-proton ratios. Experimenters can create such nuclei using artificially induced fusion or nucleon transfer reactions, employing ion beams from a particle accelerator. Beams with even higher energies can be used to create nuclei at very high temperatures, and there are signs that these experiments have produced a phase transition from normal nuclear matter to a new state, the quark-gluon plasma, in which quarks mingle with one another, rather than being segregated in triplets as they are in neutrons and protons.

Modern Topics in Nuclear Physics

Nuclear Decay: Spontaneous Changes from one Nuclide to another

There are 80 elements that have at least one stable isotope, and 250 such stable isotopes. However, there are thousands more well-characterized isotopes that are unstable. These radioisotopes may be unstable and decay along timescales ranging from fractions of a second to weeks, years, or even many millions of years.

If a nucleus has too few or too many neutrons it may be unstable, and will decay after some period of time. For example, in a process called beta decay, a nitrogen-16 atom (7 protons, 9 neutrons) is converted to an oxygen-16 atom (8 protons, 8 neutrons) within a few seconds of being created. In this decay, a neutron in the nitrogen nucleus is turned into a proton and an electron and antineutrino, by the weak nuclear force. The element is transmuted to another element in the process because, while it previously had seven protons (which makes it nitrogen), it now has eight (which makes it oxygen).

In alpha decay, the radioactive element decays by emitting a helium nucleus (2 protons and

2 neutrons), giving another element, plus helium-4. In many cases this process continues through several steps of this kind, including other types of decays, until a stable element is formed.

In gamma decay, a nucleus decays from an excited state into a lower energy state by emitting a gamma ray. The element is not changed in the process.

Other, more exotic decays, are also possible. For example, in internal conversion decay, the energy from an excited nucleus may be used to eject one of the inner orbital electrons from the atom. This process produces high speed electrons, but it is not beta decay, and (unlike beta decay) does not transmute one element to another.

Nuclear Fusion

When two light nuclei come into very close contact with each other, it is possible for the strong force to fuse the two together. It takes a great deal of energy to push the nuclei close enough together for the strong or nuclear forces to have an effect, so the process of nuclear fusion can take place only at very high temperatures or high densities. Once the nuclei are close enough together, the strong force overcomes their electromagnetic repulsion and squishes them into a new nucleus. A very large amount of energy is released when light nuclei fuse together because the binding energy per nucleon increases with mass number up until nickel-62.

Stars like the Sun are powered by the fusion of four protons into a helium nucleus, two positrons, and two neutrinos. The uncontrolled fusion of hydrogen into helium is known as thermonuclear runaway. Research to find an economically viable method of using energy from a controlled fusion reaction is currently being undertaken by various research establishments.

Nuclear Fission

For nuclei heavier than nickel-62, the binding energy per nucleon decreases with mass number. It is therefore possible for energy to be released if a heavy nucleus breaks apart into two lighter ones. This splitting of atomic nuclei is known as nuclear fission.

The process of alpha decay may be thought of as a special type of spontaneous nuclear fission. This process produces a highly asymmetrical fission because the four particles that make up the alpha particle are especially tightly bound to each other, making production of this nucleus in fission particularly likely.

For some of the heaviest nuclei that produce neutrons on fission, and which also easily absorb neutrons to initiate fission, a self-igniting type of neutron-initiated fission can be obtained, in a so-called chain reaction. Chain reactions were known in chemistry before physics, and in fact many familiar processes like fires and chemical explosions are chemical chain reactions.

The fission or "nuclear" chain-reaction, using fission-produced neutrons, is the source of energy for nuclear power plants and fission type nuclear bombs such as the two that the United States used against Hiroshima and Nagasaki to bring an end to World War II in the Pacific theater. Heavy nuclei such as uranium and thorium may undergo spontaneous fission, but they are much more likely to undergo alpha decay.

For a neutron-initiated chain-reaction to occur, there must be a critical mass of the element present in a certain space under certain conditions (these conditions slow and conserve neutrons for the reactions). There is one known example of a natural nuclear fission reactor, which was active in two regions of Oklo, Gabon, Africa, over 1.5 billion years ago. Measurements of natural neutrino emission have demonstrated that around half of the heat emanating from the Earth's core results from radioactive decay. However, it is not known if any of this results from fission chain-reactions.

Production of Heavy Elements

As the Universe cooled after the Big Bang, it eventually became possible for particles as we know them to exist. The most common particles created in the Big Bang that are still easily observable to us were protons (hydrogen) and electrons (in equal numbers). Some heavier elements were created as the protons collided with each other, but most of the heavy elements we see today were created inside of stars during a series of fusion stages, such as the proton-proton chain, the CNO cycle, and the triple-alpha process.

Progressively heavier elements are created during the evolution of a star. Since the binding energy per nucleon peaks around iron, energy is released only through fusion processes occurring below this point. Since the creation of heavier nuclei by fusion costs energy, nature resorts to the process of neutron capture. Neutrons (due to their lack of charge) are readily absorbed by a nucleus. The heavy elements are created by either a slow neutron capture process (the so-called s process) or by the rapid (or r) process. The s process occurs in thermally pulsing stars (called AGB, or asymptotic giant branch stars) and takes hundreds to thousands of years to reach the heaviest elements of lead and bismuth. The r process is thought to occur in supernova explosions because the conditions of high temperature, high neutron flux and ejected matter are present. These stellar conditions make the successive neutron captures very fast, involving very neutron-rich species, which then undergo beta decay to heavier elements, especially at the so-called waiting points that correspond to more stable nuclides with closed neutron shells (magic numbers). The r process duration is typically in the range of a few seconds.

Nuclear Chemistry

Nuclear chemistry is the study of the chemical and physical properties of elements as influenced by changes in the structure of the atomic nucleus. Modern nuclear chemistry, sometimes referred to as radiochemistry, has become very interdisciplinary in its applications, ranging from the study of the formation of the elements in the universe to the design of radioactive drugs for diagnostic medicine. In fact, the chemical techniques pioneered by nuclear chemists have become so important that biologists, geologists, and physicists use nuclear chemistry as ordinary tools of their disciplines. While the common perception is that nuclear chemistry involves only the study of radioactive nuclei, advances in modern mass spectrometry instrumentation has made chemical studies using stable, nonradioactive isotopes increasingly important.

There are essentially three sources of radioactive elements. Primordial nuclides are radioactive elements whose half-lives are comparable to the age of our solar system and were present at the formation of Earth. These nuclides are generally referred to as naturally occurring radioactivity and are derived from the radioactive decay of thorium and uranium. Cosmogenic nuclides are atoms that are constantly being synthesized from the bombardment of planetary surfaces by cosmic particles (primarily protons ejected from the Sun), and are also considered natural in their origin. The third source of radioactive nuclides is termed anthropogenic and results from human activity in the production of nuclear power, nuclear weapons, or through the use of particle accelerators. Lasers focus on a small pellet of fuel in attempt to create a nuclear fusion reaction (the combination of two nuclei to produce another nucleus) for the purpose of producing energy.

Marie Curie was the founder of the field of nuclear chemistry. She was fascinated by Antoine-Henri Becquerel's discovery that uranium minerals can emit rays that are able to expose photographic film, even if the mineral is wrapped in black paper. Using an electrometer invented by her husband Pierre and his brother Jacques that measured the electrical conductivity of air (a precursor to the Geiger counter), she was able to show that thorium also produced these rays—a process that she called radioactivity. Through tedious chemical separation procedures involving precipitation of different chemical fractions, Marie was able to show that a separated fraction that had the chemical properties of bismuth and another fraction that had the chemical properties of barium were much more radioactive per unit mass than the original uranium ore. She had separated and discovered the elements polonium and radium, respectively. Further purification of radium from barium produced approximately 100 milligrams of radium from an initial sample of nearly 2,000 kilograms of uranium ore.

In 1911 Ernest Rutherford asked a student, George de Hevesy, to separate a lead impurity from a decay product of uranium, radium-D. De Hevesy did not succeed in this task (we now know that radium-D is the radioactive isotope 210 Pb), but this failure gave rise to the idea of using radioactive isotopes as tracers of chemical processes. With Friedrich Paneth in Vienna in 1913, de Hevesy used 210 Pb to measure the solubility of lead salts—the first application of an isotopic tracer technique. De Hevesy went on to pioneer the application of isotopic tracers to study biological processes and is generally considered to be the founder of a very important area in which nuclear chemists work today, the field of nuclear medicine. De Hevesy also is credited with discovering the technique of neutron activation analysis, in which samples are bombarded by neutrons in a nuclear reactor or from a neutron generator, and the resulting radioactive isotopes are measured, allowing the analysis of the elemental composition of the sample.

Lasers focus on a small pellet of fuel in attempt to create a nuclear fusion reaction (the combination
of two nuclei to produce another nucleus) for the purpose of producing energy.

In Germany in 1938, Otto Hahn and Fritz Strassmann, skeptical of claims by Enrico Fermi and Irène
Joliot-Curie that bombardment of uranium by neutrons produced new so-called transuranic elements
(elements beyond uranium), repeated these experiments and chemically isolated a radioactive isotope
of barium. Unable to interpret these findings, Hahn asked Lise Meitner, a physicist and former col-
league, to propose an explanation for his observations. Meitner and her nephew, Otto Frisch, showed
that it was possible for the uranium nucleus to be split into two smaller nuclei by the neutrons, a process that they termed " fission ." The discovery of nuclear fission eventually led to the development of
nuclear weapons and, after World War II, the advent of nuclear power to generate electricity. Nuclear chemists were involved in the chemical purification of plutonium obtained from uranium targets
that had been irradiated in reactors. They also developed chemical separation techniques to isolate
radioactive isotopes for industrial and medical uses from the fission products wastes associated with
plutonium production for weapons. Today, many of these same chemical separation techniques are
being used by nuclear chemists to clean up radioactive wastes resulting from the fifty-year production
of nuclear weapons and to treat wastes derived from the production of nuclear power.

In 1940, at the University of California in Berkeley, Edwin McMillan and Philip Abelson produced the
first manmade element, neptunium (Np), by the bombardment of uranium with low energy neutrons
from a nuclear accelerator. Shortly thereafter, Glenn Seaborg, Joseph Kennedy, Arthur Wahl, and
McMillan made the element plutonium by bombarding uranium targets with deuterons, particles
derived from the heavy isotope of hydrogen, deuterium (^{2}H). Both McMillan and Seaborg recognized
that the chemical properties of neptunium and plutonium did not resemble those of rhenium and
osmium, as many had predicted, but more closely resembled the chemistry of uranium, a fact that led
Seaborg in 1944 to propose that the transuranic elements were part of a new group of elements called
the actinide series that should be placed below the lanthanide series on the periodic chart. Seaborg
and coworkers went on to discover many more new elements and radioactive isotopes and to study
their chemical and physical properties. At the present, nuclear chemists are involved in trying to discover new elements beyond the 112 that are presently confirmed and to study the chemical properties
of these new elements, even though they may exist for only a few thousandths of a second.

As early as 1907 Bertram Boltwood had used the discovery of radioactive decay laws by Ernest
Rutherford and Frederick Soddy to ascribe an age of over two billion years to a uranium mineral.
In 1947 Willard Libby at the University of Chicago used the decay of ^{14}C to measure the age of

dead organic matter. The cosmogenic radionuclide, ^{14}C, becomes part of all living matter through photosynthesis and the consumption of plant matter. Once the living organism dies, the ^{14}C decays at a known rate, enabling a date for the carbon-containing relic to be calculated. Today, scientists ranging from astrophysicists to marine biologists use the principles of radiometric dating to study problems as diverse as determining the age of the universe to defining food chains in the oceans. In addition, newly developed analytical techniques such as accelerator mass spectrometry (AMS) have allowed nuclear chemists to extend the principles of radiometric dating to nonradioactive isotopes in order to study modern and ancient processes that are affected by isotopic fractionation. This isotopic fractionation results from temperature differences in the environment in which the material was formed (at a given temperature, the lighter isotope will be very slightly more reactive than the heavier isotope), or from different chemical reaction sequences.

The newest area in which nuclear chemists play an important role is the field of nuclear medicine. Nuclear medicine is a rapidly expanding branch of health care that uses short-lived radioactive isotopes to diagnose illnesses and to treat specific diseases. Nuclear chemists synthesize drugs from radionuclides produced in nuclear reactors or accelerators that are injected into the patient and will then seek out specific organs or cancerous tumors. Diagnosis involves use of the radiopharmaceutical to generate an image of the tumor or organ to identify problems that may be missed by x rays or physical examinations. Treatment involves using radioactive compounds at carefully controlled doses to destroy tumors. These nuclear medicine techniques hold much promise for the future because they use biological chemistry to specify target cells much more precisely than traditional radiation therapy, which uses radiation from external sources to kill tumor cells, killing nontarget cells as well. Additionally, the use of nuclear pharmaceuticals containing the short-lived isotope ^{11}C has allowed nuclear chemists and physicians to probe brain activity to better understand the biochemical basis of illnesses ranging from Parkinson's disease to drug abuse.

Nuclear Reaction

A nuclear reaction is considered to be the process in which two nuclear particles (two nuclei or a nucleus and a nucleon) interact to produce two or more nuclear particles or γ-rays (gamma rays). Thus, a nuclear reaction must cause a transformation of at least one nuclide to another. Sometimes if a nucleus interacts with another nucleus or particle without changing the nature of any nuclide, the process is referred to a nuclear scattering, rather than a nuclear reaction. Perhaps the most notable nuclear reactions are the nuclear fusion reactions of light elements that power the energy production of stars and the Sun. Natural nuclear reactions occur also in the interaction between cosmic rays and matter.

The most notable man-controlled nuclear reaction is the fission reaction which occurs in nuclear reactors. Nuclear reactors are devices to initiate and control a nuclear chain reaction, but there are not only manmade devices. The world's first nuclear reactor operated about two billion years ago. The natural nuclear reactor formed at Oklo in Gabon, Africa, when a uranium-rich mineral deposit became flooded with groundwater that acted as a neutron moderator, and a nuclear chain reaction started. These fission reactions were sustained for hundreds of thousands of years, until a chain reaction could no longer be supported. This was confirmed by existence of isotopes of the fission-product gas xenon and by different ratio of U-235/U-238 (enrichment of natural uranium).

Notation of Nuclear Reactions

Standard nuclear notation shows the chemical symbol, the mass number and the atomic number of the isotope.

If the initial nuclei are denoted by a and b, and the product nuclei are denoted by c and d, the reaction can be represented by the equation:

$$a + b \rightarrow c + d$$

$$^{10}_{5}B + \, ^{7}_{3}Li + \, ^{4}_{2}He + 2.8 \, MeV$$

Instead of using the full equations in the style above, in many situations a compact notation is used to describe nuclear reactions. This style of the form a(b,c)d is equivalent to a + b producing c + d. Light particles are often abbreviated in this shorthand, typically p means proton, n means neutron, d means deuteron, α means an alpha particle or helium-4, β means beta particle or electron, γ means gamma photon, etc. The reaction above would be written as 10B(n,α)7Li.

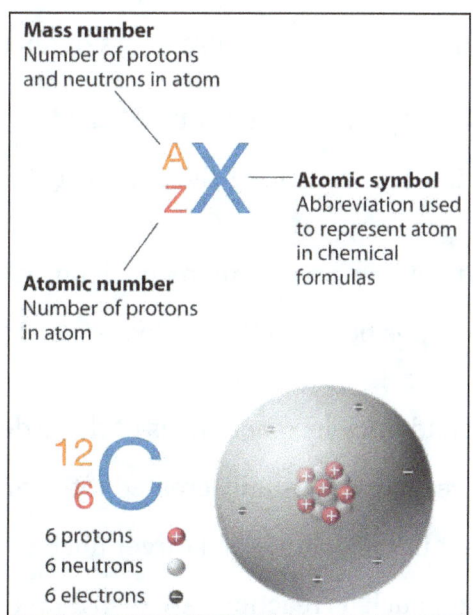

Basic Classification of Nuclear Reactions

In order to understand the nature of neutron nuclear reactions, the classification according to the time scale of of these reactions has to be introduced. Interaction time is critical for defining the reaction mechanism.

There are two extreme scenarios for nuclear reactions (not only neutron reactions):

- A projectile and a target nucleus are within the range of nuclear forces for the very short time allowing for an interaction of a single nucleon only. These type of reactions are called the direct reactions.

- A projectile and a target nucleus are within the range of nuclear forces for the time allowing for a large number of interactions between nucleons. These type of reactions are called the compound nucleus reactions.

In fact, there is always some non-direct (multiple internuclear interaction) component in all reactions, but the direct reactions have this component limited.

Basic Characteristics of Direct Reactions

- The direct reactions are fast and involve a single-nucleon interaction.
- The interaction time must be very short ($\sim 10^{-22}$ s).
- The direct reactions require incident particle energy larger than $\sim$ 5 MeV/Ap. (Ap is the atomic mass number of a projectile).
- Incident particles interact on the surface of a target nucleus rather than in the volume of a target nucleus.
- Products of the direct reactions are not distributed isotropically in angle, but they are forward focused.
- Direct reactions are of importance in measurements of nuclear structure.

Basic Characteristics of Compound Nucleus Reactions

- The compound nucleus is a relatively long-lived intermediate state of particle-target composite system.
- The compound nucleus reactions involve many nucleon-nucleon interactions.
- The large number of collisions between the nucleons leads to a thermal equilibrium inside the compound nucleus.
- The time scale of compound nucleus reactions is of the order of 10^{-18} s $-$ 10^{-16} s.
- The compound nucleus reactions is usually created if the projectile has low energy.
- Incident particles interact in the volume of a target nucleus.
- Products of the compound nucleus reactions are distributed near isotropically in angle (the nucleus loses memory of how it was created – the Bohr's hypothesis of independence).
- The mode of decay of compound nucleus do not depend on the way the compound nucleus is formed.
- Resonances in the cross-section are typical for the compound nucleus reaction.

Types of Nuclear Reactions

Although the number of possible nuclear reactions is enormous, nuclear reactions can be sorted by types. Most of nuclear reactions are accompanied by gamma emission. Some examples are:

- Elastic scattering - Occurs, when no energy is transferred between the target nucleus and the incident particle.

208Pb (n, n) 208Pb

- Inelastic scattering - Occurs, when energy is transferred. The difference of kinetic energies is saved in excited nuclide.

40Ca (α, α') 40mCa

- Capture reactions - Both charged and neutral particles can be captured by nuclei. This is accompanied by the emission of γ-rays. Neutron capture reaction produces radioactive nuclides (induced radioactivity).

238U (n, γ) 239U

- Transfer Reactions - The absorption of a particle accompanied by the emission of one or more particles is called the transfer reaction.

4He (α, p) 7Li

Fission reactions - Nuclear fission is a nuclear reaction in which the nucleus of an atom splits into smaller parts (lighter nuclei). The fission process often produces free neutrons and photons (in the form of gamma rays), and releases a large amount of energy.

235U (n, 3 n) fission products

- Fusion reactions - Occur when, two or more atomic nuclei collide at a very high speed and join to form a new type of atomic nucleus. The fusion reaction of deuterium and tritium is particularly interesting because of its potential of providing energy for the future.

3T (d, n) 4He

- Spallation reactions - Occur, when a nucleus is hit by a particle with sufficient energy and momentum to knock out several small fragments or, smash it into many fragments.

- Nuclear decay (Radioactive decay) - Occurs when an unstable atom loses energy by emitting ionizing radiation. Radioactive decay is a random process at the level of single atoms, in that, according to quantum theory, it is impossible to predict when a particular atom will decay. There are many types of radioactive decay:

  ○ Alpha radioactivity - Alha particles consist of two protons and two neutrons bound together into a particle identical to a helium nucleus. Because of its very large mass (more than 7000 times the mass of the beta particle) and its charge, it heavy ionizes material and has a very short range.

  ○ Beta radioactivity - Beta particles are high-energy, high-speed electrons or positrons emitted by certain types of radioactive nuclei such as potassium-40. The beta particles have greater range of penetration than alpha particles, but still much less than gamma rays. The beta particles emitted are a form of ionizing radiation also known as beta rays. The production of beta particles is termed beta decay.

  ○ Gamma radioactivity - Gamma rays are electromagnetic radiation of an very high frequency and are therefore high energy photons. They are produced by the decay of nuclei as they transition from a high energy state to a lower state known as gamma decay. Most of nuclear reactions are accompanied by gamma emission.

◦ Neutron emission - Neutron emission is a type of radioactive decay of nuclei containing excess neutrons (especially fission products), in which a neutron is simply ejected from the nucleus. This type of radiation plays key role in nuclear reactor control, because these neutrons are delayed neutrons.

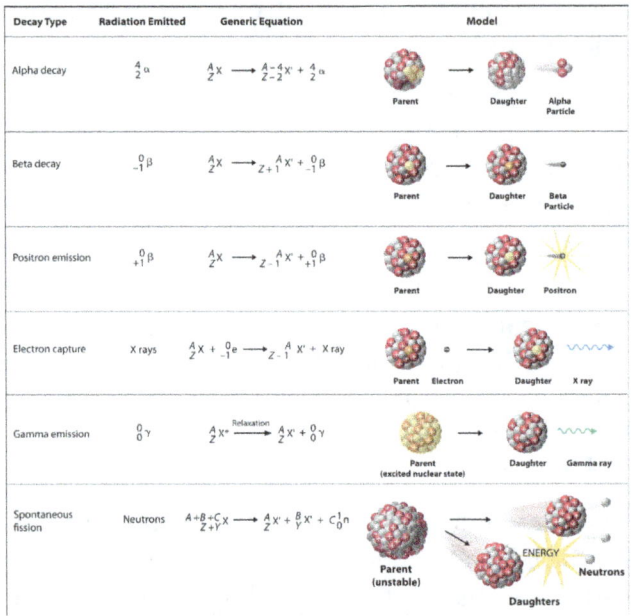

Conservation Laws in Nuclear Reactions

In analyzing nuclear reactions, we apply the many conservation laws. Nuclear reactions are subject to classical conservation laws for charge, momentum, angular momentum, and energy (including rest energies). Additional conservation laws, not anticipated by classical physics, are:

Law of Conservation of Lepton Number

Q/e	$L_e = -1$	$L_\mu = -1$	$L_T = -1$
0	$\overline{v}_e$	$\overline{v}_\mu$	$\overline{v}_\tau$
+1	e^+	μ^+	τ^+
Q/e	$L_e = 1$	$L_\mu = 1$	$L_T = 1$
0	$\overline{v}_e$	v_μ	v_τ
-1	e^-	μ^-	τ^-

In particle physics, the lepton number is used to denote which particles are leptons and which particles are not. Each lepton has a lepton number of 1 and each antilepton has a lepton number of -1. Other non-leptonic particles have a lepton number of 0. The lepton number is a conserved quantum number in all particle reactions. A slight asymmetry in the laws of physics allowed leptons to be created in the Big Bang.

The conservation of lepton number means that whenever a lepton of a certain generation is created or destroyed in a reaction, a corresponding antilepton from the same generation must be created or destroyed. It must be added, there is a separate requirement for each of the three generations of leptons, the electron, muon and tau and their associated neutrinos.

Consider the decay of the neutron. The reaction involves only first generation leptons: electrons and neutrinos:

$$n \rightarrow p + e^- + \overline{v_e}$$

Lepton number: 0→0+1

Since the lepton number must be equal to zero on both sides and it was found that the reaction is a three-particle decay (the electrons emitted in beta decay have a continuous rather than a discrete spectrum), the third particle must be an electron antineutrino.

Law of Conservation of Baryon Number

n particle physics, the baryon number is used to denote which particles are baryons and which particles are not. Each baryon has a baryon number of 1 and each antibaryonhas a baryon number of -1. Other non-baryonic particles have a baryon number of 0. Since there are exotic hadrons like pentaquarks and tetraquarks, there is a general definition of baryon number as:

$$B = \frac{1}{3}(n_q - n_{\bar{q}})$$

Where, n_q is the number of quarks, and nq is the number of antiquarks.

The baryon number is a conserved quantum number in all particle reactions. The law of conservation of baryon number states that:

The sum of the baryon number of all incoming particles is the same as the sum of the baryon numbers of all particles resulting from the reaction.

For example, the following reaction has never been observed:

$$p + n \rightarrow p + \overline{p}$$

Baryon number: 1+1→1+1-1

If the incoming proton has sufficient energy and charge, energy, and so on, are conserved. This reaction does not conserve baryon number since the left side has B =+2, and the right has B =+1.

On the other hand, the following reaction (proton-antiproton pair production) does conserve B and does occur if the incoming proton has sufficient energy (the threshold energy = 5.6 GeV):

$$p + p \rightarrow p + p + \overline{p} + p$$

Baryon number: 1+1→1+1-1+1

As indicated, B = +2 on both sides of this equation.

From these and other reactions, the conservation of baryon number has been established as a basic principle of physics.

This principle provides basis for the stability of the proton. Since the proton is the lightest particle among all baryons, the hypothetical products of its decay would have to be non-baryons. Thus, the decay would violate the conservation of baryon number. It must be added some theories have suggested that protons are in fact unstable with very long half-life ($\sim 10^{30}$ years) and that they decay into leptons. There is currently no experimental evidence that proton decay occurs.

Law of Conservation of Electric Charge

The law of conservation of electric charge can be demonstrated also on positron-electron pair production. Since a gamma ray is electrically neutral and sum of the electric charges of electron and positron is also zero, the electric charge in this reaction is also conserved.

$$\gamma \rightarrow e^- + e^+$$

It must be added, in order for electron-positron pair production to occur, the electromagnetic energy of the photon must be above a threshold energy, which is equivalent to the rest mass of two electrons. The threshold energy (the total rest mass of produced particles) for electron-positron pair production is equal to 1.02MeV (2×0.511MeV) because the rest mass of a single electron is equivalent to 0.511MeV of energy. If the original photon's energy is greater than 1.02MeV, any energy above 1.02MeV is according to the conservation law split between the kinetic energy of motion of the two particles. The presence of an electric field of a heavy atom such as lead or uranium is essential in order to satisfy conservation of momentum and energy. In order to satisfy both conservation of momentum and energy, the atomic nucleus must receive some momentum. Therefore a photon pair production in free space cannot occur.

Certain of these laws are obeyed under all circumstances, others are not. We have accepted conservation of energy and momentum. In all the examples given we assume that the number of protons and the number of neutrons is separately conserved. We shall find circumstances and conditions in which this rule is not true. Where we are considering non-relativistic nuclear reactions, it is essentially true. However, where we are considering relativistic nuclear energies or those involving the weak interactions, we shall find that these principles must be extended.

Some conservation principles have arisen from theoretical considerations, others are just empirical relationships. Notwithstanding, any reaction not expressly forbidden by the conservation laws will generally occur, if perhaps at a slow rate. This expectation is based on quantum mechanics. Unless the barrier between the initial and final states is infinitely high, there is always a non-zero probability that a system will make the transition between them.

For purposes of analyzing non-relativistic reactions, it is sufficient to note four of the fundamental laws governing these reactions.

- Conservation of nucleons - The total number of nucleons before and after a reaction are the same.

- Conservation of charge - The sum of the charges on all the particles before and after a reaction are the same.

- Conservation of momentum - The total momentum of the interacting particles before and after a reaction are the same.

- Conservation of energy - Energy, including rest mass energy, is conserved in nuclear reactions.

Energetics of Nuclear Reactions: Q-value

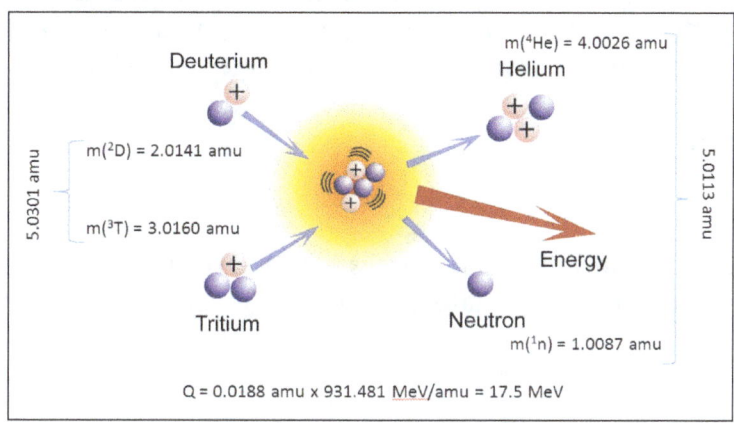

Q-value of DT fusion reaction.

In nuclear and particle physics the energetics of nuclear reactions is determined by the Q-value of that reaction. The Q-value of the reaction is defined as the difference between the sum of the masses of the initial reactants and the sum of the masses of the final products, in energy units (usually in MeV).

Consider a typical reaction, in which the projectile a and the target A gives place to two products, B and b. This can also be expressed in the notation that we used so far, $a + A \rightarrow B + b$, or even in a more compact notation, $A(a,b)B$.

The Q-value of this reaction is given by:

$$Q = [m_a + m_A - (m_b + m_B)]c^2$$

which is the same as the excess kinetic energy of the final products:

$$Q = T_{final} - T_{initial}$$
$$= T_b + T_B - (T_a + T_A)$$

For reactions in which there is an increase in the kinetic energy of the products Q is positive. The positive Q reactions are said to be exothermic (or exergic). There is a net release of energy, since the kinetic energy of the final state is greater than the kinetic energy of the initial state.

For reactions in which there is a decrease in the kinetic energy of the products Q is negative. The negative Q reactions are said to be endothermic (or endoergic) and they require a net energy input.

The energy released in a nuclear reaction can appear mainly in one of three ways:

- Kinetic energy of the products.

- Emission of gamma rays: Gamma rays are emitted by unstable nuclei in their transition from a high energy state to a lower state known as gamma decay.

- Metastable state: Some energy may remain in the nucleus, as a metastable energy level.

A small amount of energy may also emerge in the form of X-rays. Generally, products of nuclear reactions may have different atomic numbers, and thus the configuration of their electron shells is different in comparison with reactants. As the electrons rearrange themselves and drop to lower energy levels, internal transition X-rays (X-rays with precisely defined emission lines) may be emitted.

Exothermic Reactions

DT Fusion

The DT fusion reaction of deuterium and tritium is particularly interesting because of its potential of providing energy for the future. Calculate the reaction Q-value.

$$3T (d, n) 4He$$

The atom masses of the reactants and products are:

- $m(^3T) = 3.0160$ amu

- $m(^2D) = 2.0141$ amu

- $m(^1n) = 1.0087$ amu

- $m(^4He) = 4.0026$ amu

Using the mass-energy equivalence, we get the Q-value of this reaction as:

$$Q = \{(3.0160+2.0141) \text{ [amu]} - (1.0087+4.0026) \text{ [amu]}\} \times 931.481 \text{ [MeV/amu]}$$

$$= 0.0188 \times 931.481 = 17.5 \text{ MeV}$$

Tritium in Reactors

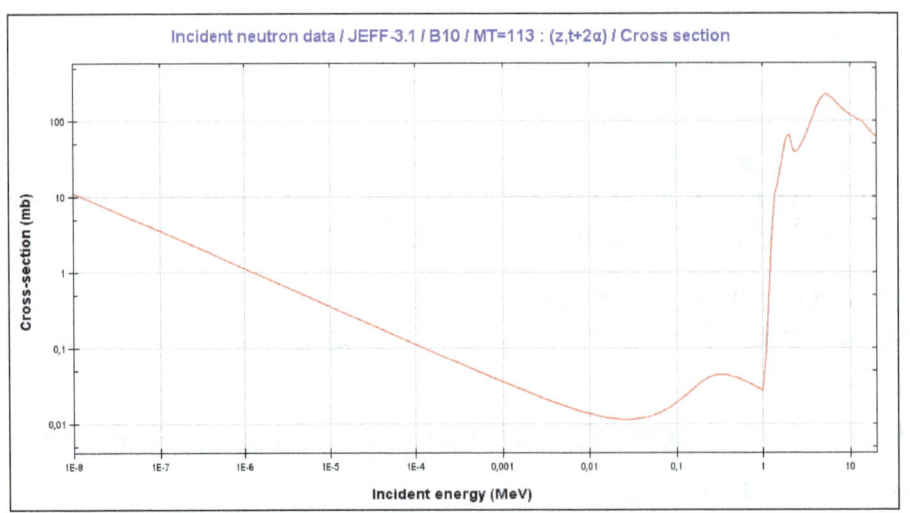

Tritium is a byproduct in nuclear reactors. Most of the tritium produced in nuclear power plants stems from the boric acid, which is commonly used as a chemical shim to compensate an excess of initial reactivity. Main reaction, in which the tritium is generated from boron is below:

10B(n,2*alpha)T

This reaction of a neutron with an isotope ^{10}B is the main way, how radioactive tritium in primary circuit of all PWRs is generated. Note that, this reaction is a threshold reaction due to its cross-section.

Calculate the reaction Q-value.

The atom masses of the reactants and products are:

- $m(^{10}B)$ = 10.01294 amu

- $m(^{1}n)$ = 1.00866 amu

- $m(^{3}T)$ = 3.01604 amu

- $m(^{4}He)$ = 4.0026 amu

Using the mass-energy equivalence, we get the Q-value of this reaction as:

$$Q = \{(10.0129+1.00866) \text{ [amu]} - (3.01604+2 \times 4.0026) \text{ [amu]}\} \times 931.481 \text{ [MeV/amu]}$$

$$= 0.00036 \times 931.481 = 0.335 \text{ MeV}$$

Endothermic Reactions

Photoneutrons

In nuclear reactors the gamma radiation plays a significant role also in reactor kinetics and in a subcriticality control. Especially in nuclear reactors with D_2O moderator (CANDU reactors) or with Be reflectors (some experimental reactors). Neutrons can be produced also in (γ, n) reactions and therefore they are usually referred to as photoneutrons.

A high energy photon (gamma ray) can under certain conditions eject a neutron from a nucleus. It occurs when its energy exceeds the binding energy of the neutron in the nucleus. Most nuclei have binding energies in excess of 6 MeV, which is above the energy of most gamma rays from fission. On the other hand there are few nuclei with sufficiently low binding energy to be of practical interest. These are: ^{2}D, ^{9}Be, ^{6}Li, ^{7}Li and ^{13}C. As can be seen from the table the lowest threshold have ^{9}Be with 1.666 MeV and ^{2}D with 2.226 MeV.

Nuclide	Threshold	Reaction
^{2}D	2.225	$^{2}H(\gamma, n)^{1}H$
^{6}Li	3.697	$^{6}Li(\gamma, n+p)^{4}He$
^{6}Li	5.67	$^{6}Li(\gamma, n)^{5}Li$
^{7}Li	7.251	$^{7}Li(\gamma, n)^{5}Li$
^{9}Be	1.667	$^{9}Be(\gamma, n)^{8}Be$
^{13}C	4.9	$^{13}C(\gamma, n)^{12}C$

In case of deuterium, neutrons can be produced by the interaction of gamma rays (with a minimum energy of 2.22 MeV) with deuterium:

$$^2_1D + \gamma \rightarrow ^1_1H + n$$

The reaction Q-value is calculated below:

The atom masses of the reactant and products are:

- $M(^2D) = 2.01363$ amu

- $m(^1n) = 1.00866$ amu

- $m(^1H) = 1.00728$ amu

Using the mass-energy equivalence, we get the Q-value of this reaction as:

$$Q = \{2.01363 \text{ [amu]} - (1.00866+1.00728) \text{ [amu]}\} \times 931.481 \text{ [MeV/amu]}$$

$$= -0.00231 \times 931.481 = -2.15 \text{ MeV}$$

(α,n) Reaction

Calculate the reaction Q-value of the following reaction:

7Li (α, n) 10B

The atom masses of the reactants and products are:

- $m(^4He) = 4.0026$ amu

- $m(^7Li) = 7.0160$ amu

- $m(^1n) = 1.0087$ amu

- $m(^{10}B) = 10.01294$ amu

Using the mass-energy equivalence, we get the Q-value of this reaction as:

$$Q = \{(7.0160+4.0026) \text{ [amu]} - (1.0087+10.01294) \text{ [amu]}\} \times 931.481 \text{ [MeV/amu]}$$

$$= 0.00304 \times 931.481 = -2.83 \text{ MeV}$$

Variance Reduction of Monte Carlo Simulation in Nuclear Engineering

The Monte Carlo method is a numerical technique that using random numbers and probability to solve problems. It represents an attempt to model nature through direct simulation for any possible results, by substituting a range of values (a probability distribution) for any factor that has inherent uncertainty.

Monte Carlo is now used routinely in many fields, such as radiation transport in the Nuclear Engineering, Dosimetry in Medical Physics field, Risk Analysis, Economics in all the applications the physical process of the solution is simulated directly based on the major components of a Monte Carlo algorithm that must be available during the simulations. The primary components of a Monte Carlo simulation are:

- Probability density functions (pdf's): The physical system must be described by a set of pdf's;

- Random number generator: A source of random numbers uniformly distributed on the unit interval must be available;

- Sampling rule: A prescription for sampling from the specified pdf's;

- Scoring: the outcomes must be accumulated into overall tallies or scores for the quantities of interest;

- Error estimation: An estimate of the statistical error (variance) as a function of the number of trials and other quantities must be determined;

- Variance reduction techniques: Methods for reducing the variance in the estimated solution to reduce the computational time for Monte Carlo simulation.

In the field of nuclear engineering, deterministic and stochastic (Monte Carlo) methods are used to solve radiation transport problems. Deterministic methods solve the transport equation for the average particle behavior and also contain uncertainties associated with the discretization of the independent variables such as space, energy and angle of the transport equation and can admit solutions that exhibit non-physical features. Although the physics of photon and electron interactions in matter is well understood, in general it is impossible to develop an analytic expression to describe particle transport in a medium. This is because the electrons can create both photons (e.g., as bremsstrahlung) and secondary or knock-on electrons (δ-rays) and conversely, photons can produce both electrons and positrons. The Monte Carlo (MC) method obtains results by simulating individual particles and recording some aspects of their average behavior. The average behavior of particles in the physical system is then inferred from the average behavior of the simulated particles. This method also enables detailed, explicit geometric, energy, and angular representations and hence is considered the most accurate method presently available for solving complex radiation transport problems. For example the most important role of Monte Carlo in radiotherapy is to obtain the dosimetric parameters with high spatial resolution. As the cost of computing in the last decades continues to decrease, applications of Monte Carlo radiation transport techniques have proliferated dramatically. On the other hand, Monte Carlo techniques have become widely used because of the availability of powerful code such as BEAM, EGSnrc, PENELOPE and ETRAN/ITS/MCNP on personal computers. These codes able to accommodate complex 3-D geometries, inclusion of flexible physics models that provide coupled electron-photon and neutron-photon transport, and the availability of extensive continuous-energy cross section libraries derived from evaluated nuclear data files.

It should be noted that these codes are general purpose, and are therefore not optimized for any particular application and are strongly depended on the solution subject. One of the difficulties associated with Monte Carlo calculations is the amount of computer time required to generate sufficient precision in the simulations. Despite substantial advancements in computational hardware performance and widespread availability of parallel computers, the computer time required for analog MC is still considered exorbitant and prohibitive for the design and analysis of many relevant real-world nuclear applications especially for the problems with complex and large geometry. But there are many ways (other than increasing simulation time) in the Monte Carlo method that users can improve the precision of the calculations. These ways known as Variance Reduction techniques and are required enabling the Monte Carlo calculation of the quantities of interest with the desired statistical uncertainty. Without the use of variance reduction techniques in complex problems, Monte Carlo code should run the problem continuously for weeks and still not obtain statistically significant reliable results. The goal of Variance Reduction techniques is to produce more accurate and precise estimate of the expected value than could be obtained in analog calculation with the same computational efforts. Variance reduction parameters are vary with problem types so iterative steps must be repeated to determine VR parameters for different problems.

Conceptual Role of the Monte Carlo Simulation

The conceptual role of the Monte Carlo simulations is to create a model similar to the real system based on known probabilities of occurrence with random sampling of the PDFs. This method is used to evaluate the average or expected behavior of a system by simulating a large number of events responsible for its behavior and observing the outcomes. Based on our experience concerning the distribution of events that occur in the system; almost any complex system can be modeled. Increasing the number of individual events (histories) improve the reported average behavior of the system.

In many applications of Monte Carlo the physical process is simulated directly and there is no need to even write down the differential equations that describe the behavior of the system. The only requirement is that the physical or mathematical system be described by probability density functions (PDF). Once the probability density functions are known the Monte Carlo simulation can proceed by random sampling from the probability density functions. Many simulations are then performed multiple trials or histories and the desired result is taken as an average over the number of observations. In many practical applications one can predict the statistical error the variance in this average result and hence an estimate of the number of Monte Carlo trials that are needed to achieve a given error.

Accuracy, Precision and Relative Error in Monte Carlo Simulation

The first component of a Monte Carlo calculation is the numerical sampling of random variables with specified PDFs. Each random variable defines as a real number that is assigned to an event. It is random because the event is random and also is variable because the assignment of the value varies over the real values. In principle, a random number is simply a particular value taken on by a random variable.

When the random number generator is used on a computer, random number sequence is not totally random. Real random numbers are hard to obtain. A logarithm function made the random number and the function repeats itself over time. When the sequence walked through, it will start from the beginning. The typical production of random numbers is in the range between 0 and 1.

A sequence of real random numbers is unpredictable and therefore un-reproducible. A random physical process, for example radioactive decay, cosmic ray arrival times, nuclear interactions, and etc, can only generate these kinds of sequences. If such a physical process is used to generate the random numbers for a Monte Carlo calculation, there is no theoretical problem. The randomness of the sequence is therefore not totally random; this phenomenon is called pseudorandom. Pseudo Random numbers look nearly random however when algorithm is not known and may be good enough for our purposes. Pesudo random numbers are generated according to a strict mathematical formula and therefore reproducible and not at all random in the mathematical sense but are supposed to be indistinguishable from a sequence generated truly randomly. That is, someone who does not know the formula is not supposed to be able to tell that a formula was used rather than a physical process. When using Monte Carlo simulation, it is desirable to have any variable depending on a uniform distributed variable, ρ. The probability, P, that a random number is smaller for a certain value, s, should be equal for both distributions.

$P (x<\rho)=P (y<s)$

The probability is in the range between 0 and 1, can be rewritten as a cumulative distribution:

$$\int_{-\infty}^{\rho} g(x)\,dx = \int_{-\infty}^{s} f(y)\,dy = \rho$$

The left side of equation $P (x<\rho)=P (y<s)$ is the uniform distribution between 0 and 1 and f(y) is the distribution needed. In this way any distribution can be made with a uniform distribution.

Monte Carlo results are obtained by simulating particle histories and assigning a score x_i to each particle history. The particle histories typically produce a range of score depending on the selected tally. By considering the f(x) as the probability density function (pdf) for selecting a particle history that scores x to the estimated tally being, the true answer (or mean) is the expected value of x, where:

$$E(X) = \int xf(x)\,dx = true\,mean$$

By assuming a scalar value for each Monte Carlo simulation output, the Monte Carlo sample mean of the first n simulation runs is defined as follow:

$$\bar{x} = \frac{1}{n}\sum_{i=1}^{n} x_i$$

Where x_i is the value of x selected from probability density function, f(x), for the ith history and n is the total number of the histories which are calculated in the problem. The sample mean $\bar{x}$, is the average value of the xi for all the histories used in the problem. But generally it does not give an accurate estimate, on the other hand there is no idea how much confidence can be considered in the estimate. So to evaluate the quantity of confidence in the estimation the sample variance can be used. Sample variance provides an estimate of how much the individual samples are spread around the mean value and is obtained as follow:

$$\delta^2 = \int \left(x - E(X)\right)^2 f(x)\,dx = E\left(x^2\right) - \left(E(x)\right)^2$$

Where δ^2 is the sample variance, E(X) is true mean and f(x) is the probability density function.

The standard deviation of scores has been obtained by the square root of the variance (δ^2), which is estimated via Monte Carlo method as s.

The standard deviation is obtained by the following equation:

$$s^2 = \frac{1}{n-1}\sum_{i=1}^{n}(x_i - \bar{x})^2 \sim \overline{x^2} - x_i^2$$

Where,

$$\overline{x^2} = \frac{1}{n}\sum_{i=1}^{n}x_i^2$$

To define the confidence interval in Monte Carlo estimation two statistical theorems are used: the law of large number and the central limit theorem.

The law of large number provides an estimate of the uncertainty in the estimate without any idea concerning the quantity of n that must be consider in calculation in practice.

To define confidence interval for the precision of a Monte Carlo result, the Central Limit Theorem of probability is used as follow:

$$\lim_{n\to\infty}\text{Prob}\left[E(x)+\alpha\frac{\delta}{\sqrt{n}}\bar{x} < E(X)+\beta\frac{\delta}{\sqrt{n}}\right] = \frac{1}{\sqrt{2\pi}}\int_{\alpha}^{\beta}e^{\frac{-t^2}{2}}dt$$

Where α and β can be any arbitrary values and n is the number of histories in the simulation. According to equation $\bar{x} = \frac{1}{n}\sum_{i=1}^{n}x_i$ as the uncertainty is proportional to $\frac{1}{\sqrt{n}}$, by increasing the number of histories by quadrupled the uncertainty in the estimation will half, which is an inherent drawback of the Monte Carlo method. So for large n, in terms of the sample standard deviation, can be rewritten as:

$$\text{Prob}\left[\alpha < \frac{\bar{x} - E(X)}{\sigma\sqrt{n}} < \beta\right] - \frac{1}{\sqrt{2\pi}}\int_{\alpha}^{\beta}e^{\frac{-t^2}{2}}dt$$

And for large n equation $\overline{x^2} = \frac{1}{n}\sum_{i=1}^{n}x_i^2$ can be written as:

$$\text{Prob}\left[\frac{\bar{x} - \lambda s_{\bar{x}}}{\sqrt{n}} \leq E(X) \leq \frac{\bar{x} + \lambda s_{\bar{x}}}{\sqrt{n}}\right] - \frac{1}{\sqrt{2\pi}}\int_{-\lambda}^{\lambda}e^{-t^2/2}dt$$

λ is the number of standard deviation, from the mean, over which the unit normal is integrated to obtain the confidence coefficient. Results for various values of λ are shown in Table. So to have confidence level the estimation for x is generally obtained as:

$$\bar{x} \pm \frac{\lambda s(x)}{\sqrt{n}}$$

For example for $\lambda=1$, the interval, $\overline{x} - \dfrac{\lambda s(x)}{\sqrt{n}}, \overline{x} + \dfrac{\lambda s(x)}{\sqrt{n}}$ has a 68% chance of containing the true mean.

Table: Results for various values of λ.

λ	0.25	0.50	1.00	1.5	2.00	3.00	4.00
Nominal Confidence Limit	20%	38%	68%	87%	95%	99%	99.99%

Equation $\lim\limits_{n\to\infty} \text{Prob}\left[E(x) + \alpha \dfrac{\delta}{\sqrt{n}} \overline{x} < E(X) + \beta \dfrac{\delta}{\sqrt{n}} \right] = \dfrac{1}{\sqrt{2\pi}} \int\limits_{\alpha}^{\beta} e^{\frac{-t^2}{2}} dt$ shows that the deviation of the sample mean from the true mean approaches zero as $n \to \infty$, and the quantity of $\delta/\sqrt{n}$ present a measured of the deviation of the sample mean from the population mean by using n samples.

To construct a confidence interval for sample mean, $\overline{x}$, that has a specified probability of the containing the true unknown mean, the sample standard deviation $s(x)$, is used to approximate the population standard deviation, $\delta(x)$. But this required that $E(x)$ and δ^2 be finite and exist.

The sample variance of $\overline{x}$ is then given by:

$$ S_{\overline{x}}^2 = \dfrac{S^2}{n} $$

It should be noted that the confidence intervals are valid only if the physical phase space is adequately sampled by the Monte Carlo calculation. The uncertainty of the Monte Carlo sampled physical phase space represents the precision of the simulation. There are several factors that can affect the precision such as tally type, variance reduction techniques and the number of histories simulated. Generally uncertainty or error caused by the statistical fluctuations of the x_i, refers to the precision of the results and not to the accuracy. Accuracy is a measure of how close the sample mean $\overline{x}$, is to the true mean.

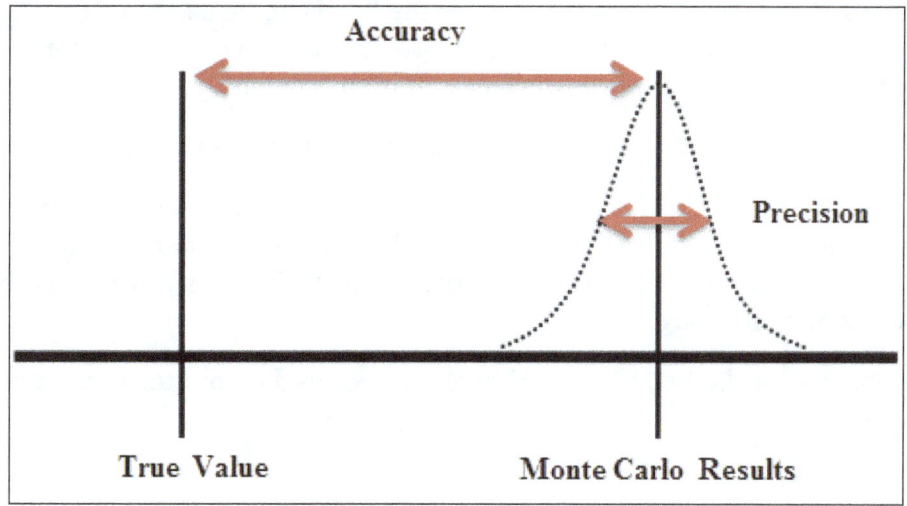

Schematic diagram of the definition for accuracy and precision.

On the other hand the difference between the true mean and the sample mean is called the systematic error. To estimate the relative error at the 1δ level which represents the statistical precision equation $\mathrm{Prob}\left[\dfrac{\overline{x}-\lambda s_{\overline{x}}}{\sqrt{n}} \le E(X) \le \dfrac{\overline{x}+\lambda s_{\overline{x}}}{\sqrt{n}}\right] - \dfrac{1}{\sqrt{2\pi}}\int\limits_{-\lambda}^{\lambda} e^{-t^{2}/2}dt$ is used.

$$R = \frac{s_{\overline{x}}}{\overline{x}}$$

In terms of Central Limit Theorem, the estimated relative error squared R^2 should be proportional to $1/n$. So as each history will take on average, the same amount of computer time and the used computer time, T, in a Monte Carlo calculation should be directly proportional to n (the number of histories); therefore $R^2 T$ should be approximately constant. Thus, the metric of efficiency for a given tally, called the figure-of-merit (FOM), includes computer time as well and define as:

$$FOM = \frac{1}{R^2 T}$$

where R is the relative error for the sample mean, and T is the total computer time taken to simulate n histories.

The FOM is also a tally reliability indicator in the sense that if the tally is well behaved, the FOM should be approximately constant (with the possible exception of statistical fluctuations early in the problem), and is thus an important and useful parameter to assess the quality (statistical behavior) of a tally bin. If the FOM is not approximately constant, the confidence intervals may not include the expected score value E(x) the expected fraction of the time.

Considering the following form of the previous relation can show the significant of the actual value of the FOM.

The above expression shows a direct relationship between computer time and the value of the FOM. Increasing the FOM for a given tally will subsequently reduce the amount of computer time required to reach a desired level of precision. Thus the FOM can be used to measure the efficiency when the variance reduction techniques have been used. The ratio of FOMs before and after using the variance reduction techniques, gives the factor of improvement.

Another use of FOM is to investigate the improvement of the new version of a Monte Carlo code.

The ratio of the FOMs for identical sample problems gives the factor of improvement. When the FOM is not a constant as a function of n, means that the result is not statistically stable; that is, no matter how many histories have been run, the important particles are showing up infrequently and have not yet been sampled enough.

Another additional use of the FOM is to estimate the required computer time to reach a desired precision by considering:

$$T \sim 1/(R^2 \times FOM)$$

Variance Reduction

The uncertainty of Monte Carlo simulation can be decreased by implementing some accurate physical models but this leads to longer calculation times. On the other hand, the accuracy of Monte Carlo dose calculation is mainly restricted by the statistical noise, because the influence of Monte Carlo method approximations should be much smaller. This statistical noise can be decreased by a larger number of histories leading to longer calculation times as well. However, there are a variety of techniques to decrease the statistical fluctuations of Monte Carlo calculations without increasing the number of particle histories. These techniques are known as variance reduction. Variance Reduction techniques are often possible to substantially decrease the relative error, R, by either producing or destroying particles, or both.

Decreasing the standard deviation, δ, and increasing the number of particle histories, n, for a given amount of computer time conflict with each other. Because decreasing δ requires more computer time per history and increasing n, results in less time per history. In general not all techniques are appropriate for all applications, also in some case some techniques tend to interfere with each other so choosing the Variance Reduction technique strongly depend on the solution.

The main goal of all the variance reduction techniques is to increase precision and decrease the relative error. The precision of the calculation is increased by increasing the number of particle histories but needs a large amount of computer running time so to accelerate the Monte Carlo simulation and reduce the computing time these techniques are applied.

Monte Carlo variance reduction techniques can be divided into four classes:

- The truncation method like geometry truncation, time and energy cut off;

- The population control method like Russian roulette, geometry splitting and weight windows;

- The modified sampling method (source biasing);

- The partially deterministic method like point detectors.

Popular Variance Reduction Techniques

Several of the more widely used variance reduction techniques are summarized as follow:

Splitting or Roulette

Geometric Splitting or Russian roulette is one of the oldest and most widely used variance reduction techniques, and when used properly, can significantly reduce the computational time of a Monte Carlo calculation.

Approximately 50% of CPU time is consumed to track secondary and higher-order photons and the electrons they set in motion. It is possible to remove a part of these photons by Russian roulette.

Generally when particles move from a region of importance I_i to a more important region I_j, ($I_i < I_j$), the particle is split into $n = I_j = I_i$ identical particles of weight w=n (if n is not an integer, splitting is done in a probabilistic manner so that the expected number of splits is equal to the importance ratio) it means that the number of particles is increased to provide better sampling and the weight of the particle is halved. Figure shows a schematic diagram of geometry splitting, when a particle moves from a lower importance region to a region with higher importance. Splitting increases the calculation time and decreases the variance whereas Russian roulette does the complete opposite.

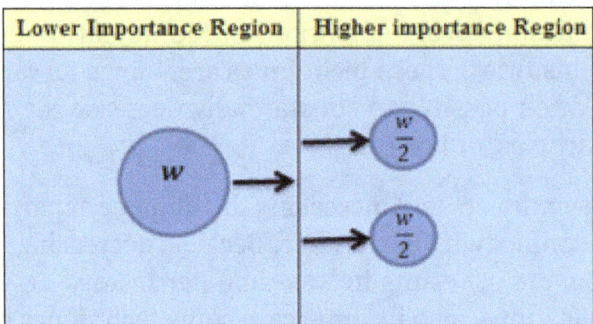

The splitting process.

In case of moving to a less important region Russian roulette is played and the particle is killed with probability 1 - (I_j=I_i), or followed further with probability I_j=I_i and weight w × I_i=I_j.

It means that the particles are killed to prevent wasting time on them.

The objective of these techniques is to spend more time sampling important spatial cells and less time sampling unimportant spatial cells. This is done by dividing the problem geometry into cells and assigning each cell i, an importance I_i.

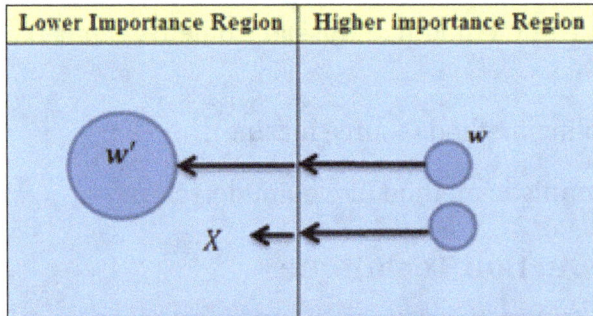

The Russian Roulette Process.

Energy splitting/roulette are similar to geometric splitting/roulette except that energy splitting/roulette is performed on the energy domain rather that on the spatial domain.

Russian roulette can be shown that the weights of all particle tracks are the same in a cell no matter which geometrical path the tracks have taken to get to the cell, assuming that no other biasing techniques, e.g. implicit capture, are used. In the simulations if a track's energy drops through a prescribed energy level, the roulette game (based on the input value of the survival probability) is played. If the game is won, the track's history is continued, but its weight is increased by the reciprocal of the survival probability to conserve weight. Russian roulette is frequently, if not always, used in radiation transport problems and can be applied at any time during the life of a particle, usually after an interaction has taken place. Russian roulette always increases variance since it cuts off histories that could still contribute to the detector, but it also always reduces the simulation time in compare with an implicit capture (which will explained later) scheme without weight thresholds.

Generally, in a deep penetration shielding problem the number of particles diminishes to almost nothing in an analog simulation, but splitting helps keep the numbers built up. To have accurate and precise results it is recommended to keep the population of tracks traveling in the desired direction more or less constant that is, approximately equal to the number of particles started from the source. Particles are killed immediately upon entering a zero importance cell, acting as a geometry cutoff. Geometry splitting/Russian roulette works well only in problems without any extreme angular dependence. In the extreme case, if no particles ever enter an important cell where the particles can be split, the Splitting/Russian roulette is useless. Energy splitting and roulette typically are used together. Energy Splitting/roulette is independent of spatial cell. If the problem has space-energy dependence, the space-energy dependent weight window is normally a better choice.

Splitting and roulette are very common techniques in Monte Carlo simulation; not only because of their simplicity but also since they only deal with variance reduction via population control and do not modify pdfs, they can be used in addition to most other techniques to have more effect.

Energy Cut Off

A Monte Carlo simulation can be made much faster, by stopping a particle once its energy drops below certain threshold energy (cutoff energy). According to the particle energy and the material that the particle is travelling through, the travelling path length of the particle can estimate. If this path length is below the required spatial resolution, particles are terminated and assume their energy is absorbed locally. This can be done by energy cut off that terminate tracks and thus decrease the time per history. Because low-energy particles can produce high energy particles, the energy cutoff can be used only when it is known that low-energy particles are either of zero or almost zero importance at the specific region (low energy particles have zero importance in some regions and high importance in others). In the Monte Carlo simulations Ecut is the photon energy cut-off parameter. It means, if a scattered photon is created with energy less than Ecut the photon will not be transported and the energy deposited locally. According to the above explanation seems the smaller Ecut the more accurate are the results but there are two criteria that should be considered in the simulations for selecting Ecut: (a) the mean free path (MFP) of photons with energy equal or less than Ecut should be small in compared with the voxel sizes or (b) the energy fraction carried by photons with energy less than Ecut is negligible compared with the energy fraction deposited. In terms of efficiency selecting the higher Ecut results in decreasing the CPU time, but on the other hand, selecting a higher value for Ecut can makes it a source of additional statistical fluctuations if it becomes comparable or even bigger than the average energy deposited by electrons. In this case the answer will be biased (low) if the energy cutoff is killing particles that might otherwise have contributed in the process even if $N \rightarrow \infty$.

Time Cut Off

A time cutoff is like a Russian roulette, with zero survival probability. The time cutoff terminates tracks and thus decreases the computer time per history. Particles are terminated when their time exceeds the time cutoff. The time cutoff can only be used in time-dependent problems where the last time bin will be earlier than the cutoff. The energy cutoff and time cutoffs are similar; but more caution must be considered with the energy cutoff because low energy particles can produce high-energy particles, whereas a late time particle cannot produce an early time particle.

Weight Window Technique

The weight window technique administers the splitting and rouletting of particles based on space and energy dependent importance. This technique is one of the most used and effective variance reduction methods that deals with both the direct decrease of variance via a large number of samples (through splitting) and the decrease of simulation time via Russian roulette, and is therefore a very effective variance reduction technique. On the other hand this technique combines Russian roulette and splitting.

To apply this variance reduction technique a lower weight bound and the width of the weight window for each energy interval of each spatial cell should be considered.

If a particle's weight is below the lower weight bound, Russian roulette is performed, and the particle's weight is either increased to be within the weight window or the particle is terminated. On the other hand, if the particle's weight is above the upper weight bound, the particle is split such that the split particles all have weights within the weight window. If the particle's weight falls within the weight window, no adjustment is performed.

As shown in figure, if a particle has a weight equal to w_{ini}, which is lower than w_L, Russian roulette will play with survival weight equal to w_s which is also provided by the user. It should be noted that the ws has to be between the windows define by w_u and w_L. If w_{ini} is greater than w_u, the particle is split into a predefine number of particles until all the particles are within the defined window. If w_{ini} is within the window the particle continues with the same weight.

One problem that may arise when using weight windows is that over-splitting might occur when a particle enters a region or it is generate in a region with higher weight than the upper limit of the weight window in that region. This can usually be solved by modifying some of the weight window parameters.

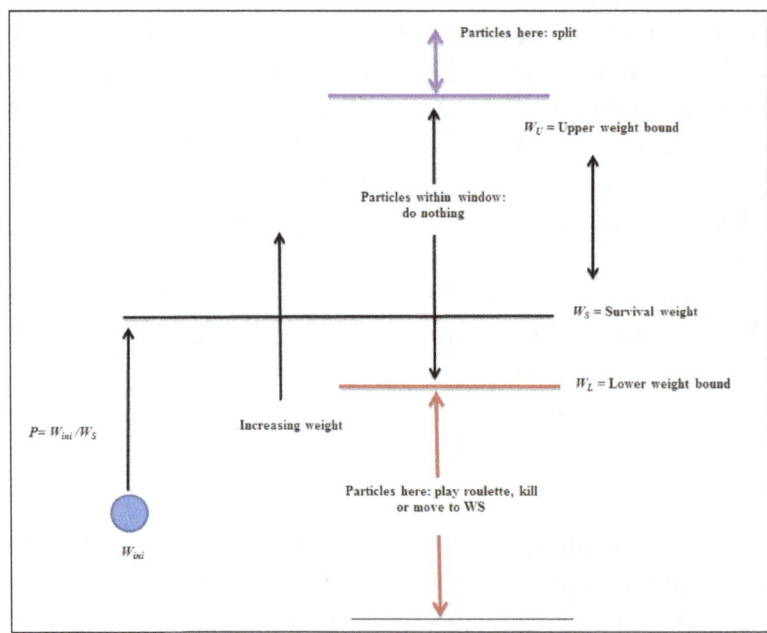

Schematic of the weight window technique.

The weight windows generator is used to determine weight windows for the simulations. When generating weight windows, it is easy to generate unwanted zeros. Zero weight windows in a region are either due to particles not entering that region or due to particles that did enter the region but did not add to the tally score. To increase the number of particles that enter or generate in a region of the system, a uniformly distributed volumetric photon source that covers the whole system is used.

The weight window generator calculates the importance of each cell in the problem. This is done by noting that the importance of a particle at a point in phase space is equal to the expected score a unit weight particle would generate. Thus, the cell's importance can be estimated as follow:

$$Importance = \frac{total\ score\ due\ to\ particles\ entering\ the\ cell}{total\ weight\ entering\ the\ cell}$$

As both the weight window and geometry splitting use the Russian roulette, there is a question concerning the difference between these two methods. The main differences are:

- The weight windows are space–energy dependent whereas geometry splitting is only dependent on space;

- The geometry splitting, splits the particles despite the weight of the particle but the weight window works completely in opposite way, it means before roulette is played and the particles split, the weight of the particle is checked against the weight window;

- The geometry splitting is only applied at surfaces, but the weight window method is applied at surfaces and collision sites or both;

- The geometry splitting method is based on the ration of importance's across the surface but weight windows utilizes absolute weights;

- As the geometry splitting is weight independent, will preserve any weight fluctuation but weight window can control weight fluctuation by other variance reduction techniques to force all particles in a cell to have a weight within a weight window.

The weight windows can be generated via the weight window generator but it requires considerable user understanding and intervention to work correctly and effectively.

Implicit Capture

Implicit capture, survival biasing, and absorption by weight reduction are synonymous. Implicit capture is a very popular variance reduction technique in radiation transport MC simulations. The implicit capture technique involves launching and tracing packets of particles instead of one by one. At launch, each packet is assigned an initial weight W_o. The packet is traced with a step length distribution determined by the total attenuation coefficient, δ. Implicit capture is a variance reduction technique that ensures that a particle always survives a collision (i.e., the particle is never absorbed).

All of the variance reduction techniques vary the physical laws of radiation transport to sample more particles in regions of the phase-space that contribute to the objective.

To compensate for this departure from the physical laws of radiation transport, the concept of particle weight, w, is introduced, where the weight can be considered as the number of particles being transported. When a variance reduction technique is applied, the weight of the particle is adjusted using the following "conservation" formula:

w (biased probability density function) = w_0 (unbiased probability density function)

where w_0 is the weight before the variance reduction technique is applied. In the implicit capture technique, a particle always survives a collision, but the particle emerges from the collision with a weight that has been reduced by a factor of δ_s/δ, which is the probability of scattering. Thus, the total particle weight is conserved.

If W_0 is the initial weight of the particle and the weight w that the particle will have after a collision, the relationship between them can be describe as follow:

$$w0 \rightarrow \begin{cases} p = \dfrac{\delta_s}{\delta} \\ p^* = 1 - p \end{cases} \; if \; W' = W \; and \; if \; W' = 0 \; \rightarrow \overline{W'} = pW_0 + p^* \times 0 = \frac{\delta_s}{\delta} W_0$$

Where, p is the probability of the particle being scattered after a collision, δ_s, is the scattering macroscopic cross-section, δ is the total macroscopic cross-section and $\overline{W'}$ is the expected outcome of the weight. For the implicit capture the particle always survives a collision with weight:

$$W' = \frac{\delta_s}{\delta} W_0$$

When implicit capture is used rather than sampling for absorption with probability $\delta_s = \delta$, the particle always survives the collision and is followed with a new weight. Implicit capture can be assumed as a splitting process in which the particle is split into absorbed weight and surviving weight. The main advantage of implicit capture is that a particle that has reached the vicinity of the tally region is not absorbed just before a score is made. In general Implicit capture always reduces the variance, but the total figure-of-merit may not improve, as the simulation time is increased because of the longer particle histories. It is, however, widely used because of its simplicity and ease of implementation.

Forced Collisions

The forced collision method is a variance reduction scheme that increases sampling of collisions in specified regions.

If the number of mean-free paths (MFP) to the next photon interaction be larger than the simulated phantom thickness, the photon will leave the region of interest without any interacting or depositing energy. Prediction of this event is not possible and to have precision results the photon behaviour must be traced through the interest region until it escapes from the region. In this case the computing time spent on the transport of escaping photons is then wasted and if the fraction of escaping photons is very large, the simulation will be very inefficient. The forced collision technique improves the efficiency by considering only the fraction of photons that interact in the phantom.

As shown in figure, when a specified particle enters a region defined as the forced collision region, the incident particle splits into un-collided and collided particles. The un-collided particle passes through the current cell without collision and is stored in the bank until later when its track is continued at the cell boundary. The collided particle is forced to collide within the specified cell.

It means that the "passing-through case" and the "collision case" are analysed due to a single incident particle simulation. These split particles are weighted by the following equation:

$$W_{uncoll} = W_0 e^{-\delta d}$$
$$W_{coll} = W_0 (1 - e^{-\delta d})$$

where W_0 is the initial weight of the particle, W_{uncoll} is the un-collided particle's weight and Wcoll is the weight of collided particle. d is the distance to region surface in the particle's direction, and δ is the macroscopic total cross section of the region material.

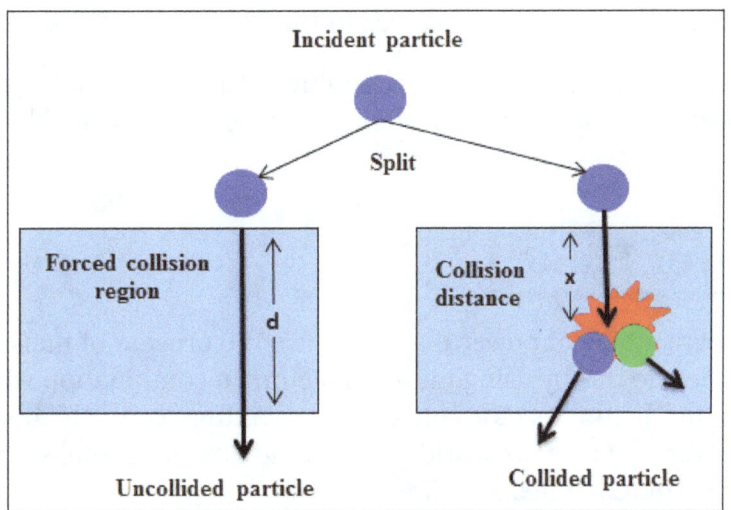

Schematic diagram of the forced collision.

The probability of colliding within a distance x is given by:

$$Prob = 1 - e^{-\delta x}$$

The particle's weight is then reduced appropriately and Russian roulette is used to ensure that the calculation time is not increased significantly by the technique.

Exponential Transformation

The exponential transform also called path length stretching is a variance reduction technique designed to enhance efficiency for deep penetration problems (e.g. shielding calculations) or surface problems (e.g. build-up in photon beams). It is often used for neutron Monte Carlo simulation and is directly applicable to photons as well.

In applying the exponential transformation a stretching parameter is used to increase distances travelled in directions of interest, while in the use of Russian roulette and splitting other parameters are introduced to increase the death probabilities in regions of low importance and the number of independent particles found in regions of high importance.

Exponential transformation samples the distance to collision from a non-analog probability density function. Specifically, it involves stretching the distance between collisions in the direction of interest and reducing the distance between collisions in directions of little interest by modifying the total macroscopic cross section as follow:

$$\delta^* = \delta(1 - p\mu)$$

Where δ^* is the modified total cross section, δ is the true total cross section, p is the exponential transform parameter used to vary the degree of biasing, $|p| < 1$, and μ is the cosine of the angle between the preferred direction and the particle's direction.

It should be mentioned that the exponential transformation technique can produce large weight fluctuations and subsequently produce unreliable mean and variance estimates. Exponential transformation generally decreases the variance but increases the time per history.

Also it should be noted that due to the large weight fluctuations that can be produced by the exponential transform the exponential transform should be used accompanied by weight control.

Nuclear Power Plant

Nuclear power plants are a type of power plant that use the process of nuclear fission in order to generate electricity. They do this by using nuclear reactors in combination with the Rankine cycle, where the heat generated by the reactor converts water into steam, which spins a turbine and a generator. Nuclear power provides the world with around 11% of its total electricity, with the largest producers being the United States and France.

The Darlington nuclear power plant in Ontario produces power from four 878 MW CANDU reactors.

Aside from the source of heat, nuclear power plants are very similar to coal-fired power plants. However, they require different safety measures since the use of nuclear fuel has vastly different properties from coal or other fossil fuels. They get their thermal power from splitting the nuclei of atoms in their reactor core, with uranium being the dominant choice of fuel in the world today. Thorium also has potential use in nuclear power production, however it is not currently in use. Below is the basic operation of a boiling water power plant, which shows the many components of a power plant, along with the generation of electricity.

Components and Operation

Nuclear Reactor

The reactor is a key component of a power plant, as it contains the fuel and its nuclear chain reaction, along with all of the nuclear waste products. The reactor is the heat source for the power plant, just like the boiler is for a coal plant. Uranium is the dominant nuclear fuel used in nuclear reactors, and its fission reactions are what produce the heat within a reactor. This heat is then transferred to the reactor's coolant, which provides heat to other parts of the nuclear power plant.

Besides their use in power generation, there are other types of nuclear reactors that are used for plutonium manufacturing, the propulsion of ships, aircraft and satellites, along with research and medical purposes. The power plant encompasses not just the reactor, but also cooling towers, turbines, generators, and various safety systems. The reactor is what makes it differ from other external heat engines.

Steam Generation

The production of steam is common among all nuclear power plants, but the way this is done varies immensely.

Steam turbine in a power plant.

The most common power plants in the world use pressurized water reactors, which use two loops of circling water to produce steam. The first loop carries extremely hot liquid water to a heat exchanger, where water at a lower pressure is circulated. It then heats up and boils to steam, and can then be sent to the turbine section.

Boiling water reactors, the second most common reactor in power generation, heat the water in the core directly to steam, as seen in figure.

Turbine and Generator

Two cooling towers of a nuclear power plant.

Once steam has been produced, it travels at high pressures and speeds through one or more turbines. These get up to extremely high speeds, causing the steam to loose energy, therefore, condensing back to a cooler liquid water. The rotation of the turbines is used to spin an electric generator, which produces electricity that is sent out the the electrical grid.

Cooling Towers

Perhaps the most iconic symbol of a nuclear power plant is the cooling towers, seen in figure. They work to reject waste heat to the atmosphere by the transfer of heat from hot water (from the turbine section) to the cooler outside air. Hot water cools in contact with the air and a small portion, around 2%, evaporates and raises up through the top. Moreover, these plants do not release any carbon dioxide—the primary greenhouse gas that contributes to climate change.

Many nuclear power plants simply put the waste heat into a river, lake or ocean instead of having cooling towers. Many other power plants like coal-fired power plants have cooling towers or these large bodies of water as well. This similarity exists because the process of turning heat into electricity is almost identical between nuclear power plants and coal-fired power plants.

Efficiency

The efficiency of a nuclear power plant is determined similarly to other heat engines—since technically the plant is a large heat engine. The amount of electric power produced for each unit of thermal power gives the plant its thermal efficiency, and due to the second law of thermodynamics there is an upper limit to how efficient these plants can be.

Typical nuclear power plants achieve efficiencies around 33-37%, comparable to fossil fueled power plants. Higher temperature and more modern designs like the Generation IV nuclear reactors could potentially reach above 45% efficiency.

References

- Nuclear-power: britannica.com, Retrieved 05 January, 2019
- Nuclear-engineering: britannica.com, Retrieved 25 May, 2019

- Materials-nuclear-engineering, nuclear-engineering: nuclear-power.net, Retrieved 23 April, 2019
- Occurrence-of-radioactivity: britannica.com, Retrieved 16 August, 2019
- Nuclear-physics: newworldencyclopedia.org, Retrieved 24 July, 2019
- Nuclear-Chemistry: chemistryexplained.com, Retrieved 16 April, 2019
- Nuclear-power-plant: energyeducation.ca, Retrieved 14 August, 2019

2

Nuclear Fusion and Fission

A type of nuclear reaction in which energy is produced by combining two nuclei is called nuclear fusion whereas a nuclear reaction in which energy is produced by splitting of a single nucleus is called nuclear fission. This chapter has been carefully written to provide an easy understanding of nuclear fusion and fission.

Nuclear Fusion

Nuclear fusion is the process by which nuclear nuclear reactions between light elements form heavier elements (up to iron). In cases where the interacting nuclei belong to elements with low atomic numbers (e.g., hydrogen [atomic number 1] or its isotopes deuterium and tritium), substantial amounts of energy are released. The vast energy potential of nuclear fusion was first exploited in thermonuclear weapons, or hydrogen bombs, which were developed in the decade immediately following World War II. Meanwhile, the potential peaceful applications of nuclear fusion, especially in view of the essentially limitless supply of fusion fuel on Earth, have encouraged an immense effort to harness this process for the production of power.

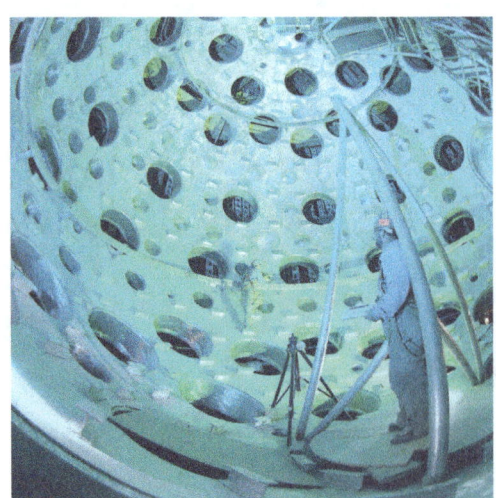

Laser-activated fusion. Interior of the U.S. Department of Energy's National Ignition Facility (NIF), The NIF target chamber uses a high-energy laser to heat fusion fuel to temperatures sufficient for thermonuclear ignition.

The Fusion Reaction

Fusion reactions constitute the fundamental energy source of stars, including the Sun. The evolution of stars can be viewed as a passage through various stages as thermonuclear reactions and nucleosynthesis cause compositional changes over long time spans. Hydrogen (H) "burning" initiates the fusion energy source of stars and leads to the formation of helium (He). Generation of fusion energy for practical use also relies on fusion reactions between the lightest elements that burn to form helium. In fact, the heavy isotopes of hydrogen—deuterium (D) and tritium (T)—react more efficiently with each other, and, when they do undergo fusion, they yield more energy per reaction than do two hydrogen nuclei. (The hydrogen nucleus consists of a single proton. The deuterium nucleus has one proton and one neutron, while tritium has one proton and two neutrons).

Fusion reactions between light elements, like fission reactions that split heavy elements, release energy because of a key feature of nuclear matter called the binding energy, which can be released through fusion or fission. The binding energy of the nucleus is a measure of the efficiency with which its constituent nucleons are bound together. Take, for example, an element with Z protons and N neutrons in its nucleus. The element's atomic weight A is Z + N, and its atomic number is Z. The binding energy B is the energy associated with the mass difference between the Z protons and N neutrons considered separately and the nucleons bound together (Z + N) in a nucleus of mass M. The formula is:

$$B = (Zm_p + Nm_n - M)C^2$$

where m_p and m_n are the proton and neutron masses and c is the speed of light. It has been determined experimentally that the binding energy per nucleon is a maximum of about 1.4×10^{-12} joule at an atomic mass number of approximately 60—that is, approximately the atomic mass number of iron. Accordingly, the fusion of elements lighter than iron or the splitting of heavier ones generally leads to a net release of energy.

Two Types of Fusion Reactions

Fusion reactions are of two basic types: (1) those that preserve the number of protons and neutrons and (2) those that involve a conversion between protons and neutrons. Reactions of the first type are most important for practical fusion energy production, whereas those of the second type are crucial to the initiation of star burning. An arbitrary element is indicated by the notation AZX, where Z is the charge of the nucleus and A is the atomic weight. An important fusion reaction for practical energy generation is that between deuterium and tritium (the D-T fusion reaction). It produces helium (He) and a neutron (n) and is written:

$$D + T \rightarrow He + n$$

To the left of the arrow (before the reaction) there are two protons and three neutrons. The same is true on the right.

The other reaction, that which initiates star burning, involves the fusion of two hydrogen nuclei to form deuterium (the H-H fusion reaction):

$$H + H \rightarrow D + \beta^+ + \nu$$

Where, β^+ represents a positron and ν stands for a neutrino. Before the reaction there are two hydrogen nuclei (that is, two protons). Afterward there are one proton and one neutron (bound together as the nucleus of deuterium) plus a positron and a neutrino (produced as a consequence of the conversion of one proton to a neutron).

Both of these fusion reactions are exoergic and so yield energy. The German-born physicist Hans Bethe proposed in the 1930s that the H-H fusion reaction could occur with a net release of energy and provide, along with subsequent reactions, the fundamental energy source sustaining the stars. However, practical energy generation requires the D-T reaction for two reasons: first, the rate of reactions between deuterium and tritium is much higher than that between protons; second, the net energy release from the D-T reaction is 40 times greater than that from the H-H reaction.

Energy Released in Fusion Reactions

Energy is released in a nuclear reaction if the total mass of the resultant particles is less than the mass of the initial reactants. To illustrate, suppose two nuclei, labeled X and a, react to form two other nuclei, Y and b, denoted:

$$X + a \rightarrow Y + b$$

The particles a and b are often nucleons, either protons or neutrons, but in general can be any nuclei. Assuming that none of the particles is internally excited (i.e., each is in its ground state), the energy quantity called the Q-value for this reaction is defined as:

$$Q = (m_x + m_a - m_b - m_y)c^2$$

where the m-letters refer to the mass of each particle and c is the speed of light. When the energy value Q is positive, the reaction is exoergic; when Q is negative, the reaction is endoergic (i.e., absorbs energy). When both the total proton number and the total neutron number are preserved before and after the reaction (as in D-T reactions), then the Q-value can be expressed in terms of the binding energy B of each particle as:

$$Q = B_y + B_b - B_x - B_a$$

The D-T fusion reaction has a positive Q-value of 2.8×10^{-12} joule. The H-H fusion reaction is also exoergic, with a Q-value of 6.7×10^{-14} joule. To develop a sense for these figures, one might consider that one metric ton (1,000 kg, or almost 2,205 pounds) of deuterium would contain roughly 3×10^{32} atoms. If one ton of deuterium were to be consumed through the fusion reaction with tritium, the energy released would be 8.4×10^{20} joules. This can be compared with the energy content of one ton of coal—namely, 2.9×10^{10} joules. In other words, one ton of deuterium has the energy equivalent of approximately 29 billion tons of coal.

Rate and Yield of Fusion Reactions

The energy yield of a reaction between nuclei and the rate of such reactions are both important. These quantities have a profound influence in scientific areas such as nuclear astrophysics and the potential for nuclear production of electrical energy.

When a particle of one type passes through a collection of particles of the same or different type, there is a measurable chance that the particles will interact. The particles may interact in many ways, such as simply scattering, which means that they change direction and exchange energy, or they may undergo a nuclear fusion reaction. The measure of the likelihood that particles will interact is called the cross section, and the magnitude of the cross section depends on the type of interaction and the state and energy of the particles. The product of the cross section and the atomic density of the target particle is called the macroscopic cross section. The inverse of the macroscopic cross section is particularly noteworthy as it gives the mean distance an incident particle will travel before interacting with a target particle; this inverse measure is called the mean free path. Cross sections are measured by producing a beam of one particle at a given energy, allowing the beam to interact with a (usually thin) target made of the same or a different material, and measuring deflections or reaction products. In this way it is possible to determine the relative likelihood of one type of fusion reaction versus another, as well as the optimal conditions for a particular reaction.

The cross sections of fusion reactions can be measured experimentally or calculated theoretically, and they have been determined for many reactions over a wide range of particle energies. They are well known for practical fusion energy applications and are reasonably well known, though with gaps, for stellar evolution. Fusion reactions between nuclei, each with a positive charge of one or more, are the most important for both practical applications and the nucleosynthesis of the light elements in the burning stages of stars. Yet, it is well known that two positively charged nuclei repel each other electrostatically—i.e., they experience a repulsive force inversely proportional to the square of the distance separating them. This repulsion is called the Coulomb barrier. It is highly unlikely that two positive nuclei will approach each other closely enough to undergo a fusion reaction unless they have sufficient energy to overcome the Coulomb barrier. As a result, the cross section for fusion reactions between charged particles is very small unless the energy of the particles is high, at least 10^4 electron volts (1 eV $\cong 1.602 \times 10^{-19}$ joule) and often more than 10^5 or 10^6 eV. This explains why the centre of a star must be hot for the fuel to burn and why fuel for practical fusion energy systems must be heated to at least 50,000,000 kelvins (K; 90,000,000 °F). Only then will a reasonable fusion reaction rate and power output be achieved.

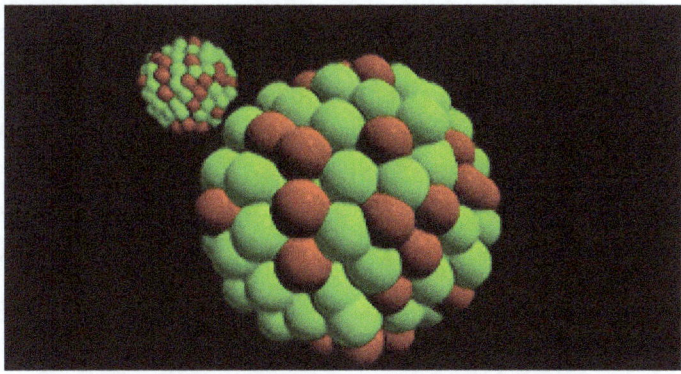

Fission of uranium nucleus Sequence of events
in the fission of a uranium nucleus by a neutron.

The phenomenon of the Coulomb barrier also explains a fundamental difference between energy generation by nuclear fusion and nuclear fission. While fission of heavy elements can be induced by either protons or neutrons, generation of fission energy for practical applications is dependent on neutrons to induce fission reactions in uranium or plutonium. Having no electric charge, the neutron is free to enter the nucleus even if its energy corresponds to room temperature. Fusion energy, relying as it does on the fusion reaction between light nuclei, occurs only when the particles are sufficiently energetic to overcome the Coulomb repulsive force. This requires the production and heating of the gaseous reactants to the high temperature state known as the plasma state.

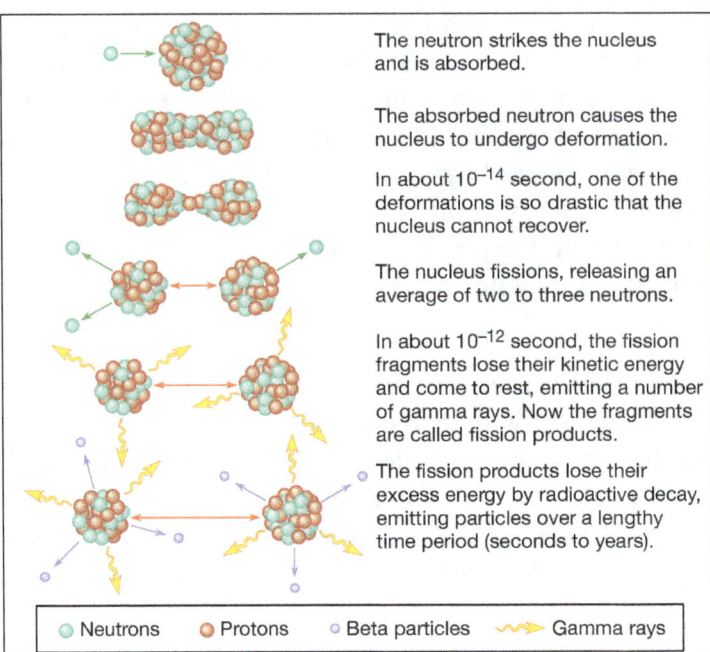

The neutron strikes the nucleus and is absorbed.

The absorbed neutron causes the nucleus to undergo deformation.

In about 10^{-14} second, one of the deformations is so drastic that the nucleus cannot recover.

The nucleus fissions, releasing an average of two to three neutrons.

In about 10^{-12} second, the fission fragments lose their kinetic energy and come to rest, emitting a number of gamma rays. Now the fragments are called fission products.

The fission products lose their excess energy by radioactive decay, emitting particles over a lengthy time period (seconds to years).

○ Neutrons   ● Protons   ○ Beta particles   〰➤ Gamma rays

Sequence of events in the fission of a uranium nucleus by a neutron.

The Plasma State

Typically, a plasma is a gas that has had some substantial portion of its constituent atoms or molecules ionized by the dissociation of one or more of their electrons. These free electrons enable plasmas to conduct electric charges, and a plasma is the only state of matter in which thermonuclear reactions can occur in a self-sustaining manner. Astrophysics and magnetic fusion research, among other fields, require extensive knowledge of how gases behave in the plasma state. The stars, the solar wind, and much of interstellar space are examples where the matter present is in the plasma state. Very high-temperature plasmas are fully ionized gases, which means that the ratio of neutral gas atoms to charged particles is small. For example, the ionization energy of hydrogen is 13.6 eV, while the average energy of a hydrogen ion in a plasma at 50,000,000 K is 6,462 eV. Thus, essentially all of the hydrogen in this plasma would be ionized.

A reaction-rate parameter more appropriate to the plasma state is obtained by accounting for the fact that the particles in a plasma, as in any gas, have a distribution of energies. That is to say, not all particles have the same energy. In simple plasmas this energy distribution is given by the Maxwell-Boltzmann distribution law, and the temperature of the gas or plasma is, within a proportionality constant, two-thirds of the average particle energy; i.e., the relationship between the average energy E and temperature T is $\overline{E} = 3kT/2$, where k is the Boltzmann constant, 8.62×10^{-5} eV per kelvin. The intensity of nuclear fusion reactions in a plasma is derived by averaging the product of the particles' speed and their cross sections over a distribution of speeds corresponding to a Maxwell-Boltzmann distribution. The cross section for the reaction depends on the energy or speed of the particles. The averaging process yields a function for a given reaction that depends only on the temperature and can be denoted $f(T)$. The rate of energy released (i.e., the power released) in a reaction between two species, a and b, is:

$$P_{ab} = n_a n_b f_{ab}(T) U_{ab},$$

where, n_a and n_b are the density of species a and b in the plasma, respectively, and U_{ab} is the energy released each time a and b undergo a fusion reaction. The parameter P_{ab} properly takes into account both the rate of a given reaction and the energy yield per reaction.

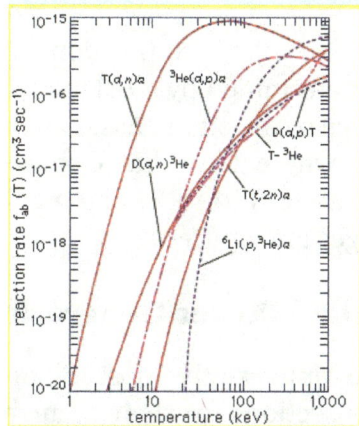

The reaction rate as a function of plasma temperature, expressed in kiloelectron volts (keV; 1 keV is equivalent to a temperature of 11,000,000 K).

Fusion Reactions in Stars

Fusion reactions are the primary energy source of stars and the mechanism for the nucleosynthesis of the light elements. In the late 1930s Hans Bethe first recognized that the fusion of hydrogen nuclei to form deuterium is exoergic (i.e., there is a net release of energy) and, together with subsequent nuclear reactions, leads to the synthesis of helium. The formation of helium is the main source of energy emitted by normal stars, such as the Sun, where the burning-core plasma has a temperature of less than 15,000,000 K. However, because the gas from which a star is formed often contains some heavier elements, notably carbon (C) and nitrogen (N), it is important to include nuclear reactions between protons and these nuclei. The reaction chain between protons that ultimately leads to helium is the proton-proton cycle. When protons also induce the burning of carbon and nitrogen, the CN cycle must be considered; and, when oxygen (O) is included, still another alternative scheme, the CNO bi-cycle, must be accounted for.

The proton-proton nuclear fusion cycle in a star containing only hydrogen begins with the reaction:

$$H + H \rightarrow D + \beta^+ + \nu; \ Q = 1.44 \text{ MeV}$$

Where, the Q-value assumes annihilation of the positron by an electron. The deuterium could react with other deuterium nuclei, but, because there is so much hydrogen, the D/H ratio is held to very low values, typically 10^{-18}. Thus, the next step is:

$$H + D \rightarrow {}^3He + \gamma; \ Q = 5.49 \text{ MeV}$$

Where, γ indicates that gamma rays carry off some of the energy yield. The burning of the helium-3 isotope then gives rise to ordinary helium and hydrogen via the last step in the chain:

$$ {}^3He + {}^3He \rightarrow {}^4He + 2(H); \ Q = 12.86 \text{ MeV}. $$

At equilibrium, helium-3 burns predominantly by reactions with itself because its reaction rate with hydrogen is small, while burning with deuterium is negligible due to the very low deuterium concentration. Once helium-4 builds up, reactions with helium-3 can lead to the production of still-heavier elements, including beryllium-7, beryllium-8, lithium-7, and boron-8, if the temperature is greater than about 10,000,000 K.

The stages of stellar evolution are the result of compositional changes over very long periods. The size of a star, on the other hand, is determined by a balance between the pressure exerted by the hot plasma and the gravitational force of the star's mass. The energy of the burning core is transported toward the surface of the star, where it is radiated at an effective temperature. The effective temperature of the Sun's surface is about 6,000 K, and significant amounts of radiation in the visible and infrared wavelength ranges are emitted.

Fusion Reactions for Controlled Power Generation

Reactions between deuterium and tritium are the most important fusion reactions for controlled power generation because the cross sections for their occurrence are high, the practical plasma temperatures required for net energy release are moderate, and the energy yield of the reactions are high—17.58 MeV for the basic D-T fusion reaction.

It should be noted that any plasma containing deuterium automatically produces some tritium and helium-3 from reactions of deuterium with other deuterium ions. Other fusion reactions involving elements with an atomic number above 2 can be used, but only with much greater difficulty. This is because the Coulomb barrier increases with increasing charge of the nuclei, leading to the requirement that the plasma temperature exceed 1,000,000,000 K if a significant rate is to be achieved. Some of the more interesting reactions are:

1. $H + {}^{11}B \rightarrow 3({}^{4}He)$; $Q = 8.68$ MeV;

2. $H + {}^{6}Li \rightarrow {}^{3}He + {}^{4}He$; $Q = 4.023$ MeV;

3. ${}^{3}He + {}^{6}Li \rightarrow H + 2({}^{4}He)$; $Q = 16.88$ MeV;

4. ${}^{3}He + {}^{6}Li \rightarrow D + {}^{7}Be$; $Q = 0.113$ MeV.

Reaction (2) converts lithium-6 to helium-3 and ordinary helium. Interestingly, if reaction (2) is followed by reaction (3), then a proton will again be produced and be available to induce reaction (2), thereby propagating the process. Unfortunately, it appears that reaction (4) is 10 times more likely to occur than reaction (3).

Methods of Achieving Fusion Energy

Practical efforts to harness fusion energy involve two basic approaches to containing a high-temperature plasma of elements that undergo nuclear fusion reactions: magnetic confinement and inertial confinement. A much less likely but nevertheless interesting approach is based on fusion catalyzed by muons; research on this topic is of intrinsic interest in nuclear physics.

Magnetic Confinement

In magnetic confinement the particles and energy of a hot plasma are held in place using magnetic fields. A charged particle in a magnetic field experiences a Lorentz force that is proportional to the product of the particle's velocity and the magnetic field. This force causes electrons and ions to spiral about the direction of the magnetic line of force, thereby confining the particles. When the topology of the magnetic field yields an effective magnetic well and the pressure balance between the plasma and the field is stable, the plasma can be confined away from material boundaries. Heat and particles are transported both along and across the field, but energy losses can be prevented in two ways. The first is to increase the strength of the magnetic field at two locations along the field line. Charged particles contained between these points can be made to reflect back and forth, an effect called magnetic mirroring. In a basically straight system with a region of intensified magnetic field at each end, particles can still escape through the ends due to scattering between particles as they approach the mirroring points. Such end losses can be avoided altogether by creating a magnetic field in the topology of a torus (i.e., configuration of a doughnut or inner tube).

External magnets can be arranged to create a magnetic field topology for stable plasma confinement, or they can be used in conjunction with magnetic fields generated by currents induced to flow in the plasma itself. The late 1960s witnessed a major advance by the Soviet Union in harnessing fusion reactions for practical energy production. Soviet scientists achieved a high plasma temperature (about 3,000,000 K), along with other physical parameters, in a machine referred to as a tokamak. A tokamak is a toroidal magnetic confinement system in which the plasma is kept stable both by an externally generated, doughnut-shaped magnetic field and by electric currents flowing within the plasma. Since the late 1960s the tokamak has been the major focus of magnetic fusion research worldwide, though other approaches such as the stellarator, the compact torus, and the reversed field pinch (RFP) have also been pursued. In these approaches, the magnetic field lines follow a helical, or screwlike, path as the lines of magnetic force proceed around the torus. In the tokamak the pitch of the helix is weak, so the field lines wind loosely around the poloidal direction (through the central hole) of the torus. In contrast, RFP field lines wind much tighter, wrapping many times in the poloidal direction before completing one loop in the toroidal direction (around the central hole).

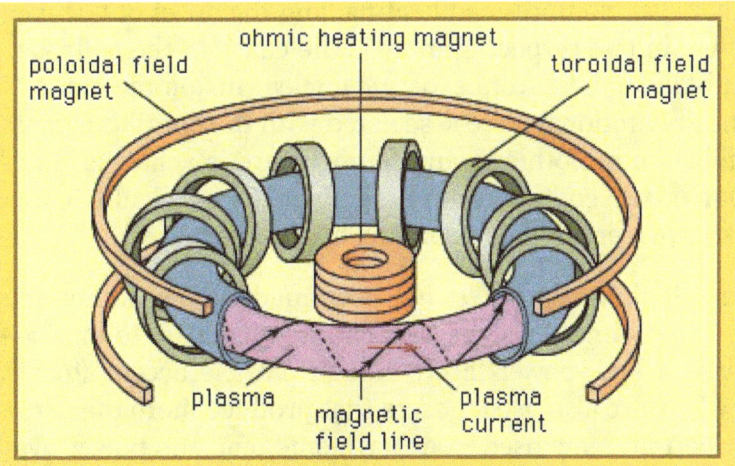

Tokamak magnetic confinement.

Magnetically confined plasma must be heated to temperatures at which nuclear fusion is vigorous, typically greater than 75,000,000 K (equivalent to an energy of 4,400 eV). This can be achieved by coupling radio-frequency waves or microwaves to the plasma particles, by injecting energetic beams of neutral atoms that become ionized and heat the plasma, by magnetically compressing the plasma, or by the ohmic heating (also known as Joule heating) that occurs when an electric current passes through the plasma.

Employing the tokamak concept, scientists and engineers in the United States, Europe, and Japan began in the mid-1980s to use large experimental tokamak devices to attain conditions of temperature, density, and energy confinement that now match those necessary for practical fusion power generation. The machines employed to achieve these results include the Joint European Torus (JET) of the European Union, the Japanese Tokamak-60 (JT-60), and, until 1997, the Tokamak Fusion Test Reactor (TFTR) in the United States. Indeed, in both the TFTR and the JET devices, experiments using deuterium and tritium produced more than 10 megawatts of fusion power and essentially energy breakeven conditions in the plasma itself. Plasma conditions approaching those achieved in tokamaks were also achieved in large stellarator machines in Germany and Japan during the 1990s.

Inertial Confinement Fusion

In this approach, a fuel mass is compressed rapidly to densities 1,000 to 10,000 times greater than normal by generating a pressure as high as 10^{17} pascals (10^{12} atmospheres) for periods as short as a nanosecond (10^{-9} second). Near the end of this time period, the implosion speed exceeds about 3×10^5 metres per second. At maximum compression of the fuel, which is now in a cool plasma state, the energy in converging shock waves is sufficient to heat the very centre of the fuel to temperatures high enough to induce fusion reactions (greater than an equivalent energy of about 4,400 eV). If the mass of this highly compressed fuel material is large enough, energy will be generated through fusion reactions before this hot plasma ball disassembles. Under proper conditions, much more energy can be released than is required to compress and shock heat the fuel to thermonuclear burning conditions.

The physical processes in ICF bear a relationship to those in thermonuclear weapons and in star formation—namely, collapse, compression heating, and the onset of nuclear fusion. The situation in star formation differs in one respect: gravity is the cause of the collapse, and a collapsed star begins to expand again due to heat from exoergic nuclear fusion reactions. The expansion is ultimately arrested by the gravitational force associated with the enormous mass of the star, at which point a state of equilibrium in both size and temperature is achieved. In contrast, the fuel in a thermonuclear weapon or ICF completely disassembles. In the ideal ICF case, however, this does not occur until about 30 percent of the fusion fuel has burned.

Over the decades, very significant progress has been made in developing the technology and systems for high-energy, short-time-pulse drivers that are necessary to implode the fusion fuel. The most common driver is a high-power laser, though particle accelerators capable of producing beams of high-energy ions are also used. Lasers that produce more than 100,000 joules in pulses of about one nanosecond are now used in experiments, and the power available in short bursts exceeds 10^{14} watts.

Two lasers capable of delivering up to 5,000,000 joules in equally short bursts, generating a power level on the fusion targets in excess of 5×10^{14} watts, are operational. One facility is the Laser Mega Joule in Bordeaux, France. The other is the National Ignition Facility at the Lawrence Livermore National Laboratory.

Muon-catalyzed Fusion

The need in traditional schemes of nuclear fusion to confine very high-temperature plasmas has led some researchers to explore alternatives that would permit fusion reactants to approach each other more closely at much lower temperatures. One method involves substituting muons (μ) for the electrons that ordinarily surround the nucleus of a fuel atom. Muons are negatively charged subatomic particles similar to electrons, except that their mass is a little more than 200 times the electron mass and they are unstable, having a half-life of about 2.2×10^{-6} second. In fact, fusion has been observed in liquid and gas mixtures of deuterium and tritium at cryogenic temperatures when muons were injected into the mixture.

Muon-catalyzed fusion is the name given to the process of achieving fusion reactions by causing a deuteron (deuterium nucleus, D^+), a triton (tritium nucleus, T^+), and a muon to form what is called a muonic molecule. Once a muonic molecule is formed, the rate of fusion reactions is approximately 3×10^{-8} second. However, the formation of a muonic molecule is complex, involving a series of atomic, molecular, and nuclear processes.

In schematic terms, when a muon enters a mixture of deuterium and tritium, the muon is first captured by one of the two hydrogen isotopes in the mixture, forming either atomic D^+-μ or T^+-μ, with the atom now in an excited state. The excited atom relaxes to the ground state through a cascade collision process, in which the muon may be transferred from a deuteron to a triton or vice versa. More important, it is also possible that a muonic molecule (D^+-μ-T^+) will be formed. Although a much rarer reaction, once a muonic molecule does form, fusion takes place almost immediately, releasing the muon in the mixture to be captured again by a deuterium or tritium nucleus and allowing the process to continue. In this sense the muon acts as a catalyst for fusion reactions within the mixture. The key to practical energy production is to generate enough fusion reactions before the muon decays.

The complexities of muon-catalyzed fusion are many and include generating the muons (at an energy expenditure of about five billion electron volts per muon) and immediately injecting them into the deuterium-tritium mixture. In order to produce more energy than what is required to initiate the process, about 300 D-T fusion reactions must take place within the half-life of a muon.

Cold Fusion and Bubble Fusion

Two disputed fusion experiments merit mention. In 1989 two chemists, Martin Fleischmann of the University of Utah and Stanley Pons of the University of Southampton in England, announced that they had produced fusion reactions at essentially room temperature. Their system consisted of electrolytic cells containing heavy water (deuterium oxide, D_2O) and palladium rods that absorbed the deuterium from the heavy water. Efforts to give a theoretical explanation of the results failed, as did worldwide efforts to reproduce the claimed cold fusion.

In 2002 Rusi Taleyarkhan and colleagues at Purdue University in Lafayette, Ind., claimed to have observed a statistically significant increase in nuclear emissions of products of fusion reactions (neutrons and tritium) during acoustic cavitation experiments with chilled deuterated (bombarded with deuterium) acetone. Their experimental setup was based on the known phenomenon of sonoluminescence. In sonoluminescence a gas bubble is imploded with high-pressure sound waves.

At the end of the implosion process, and for a short time afterward, conditions of high density and temperature are achieved that lead to light emission. By starting with larger, millimetre-sized cavitations (bubbles) that had been deuterated in the acetone liquid, the researchers claimed to have produced densities and temperatures sufficient to induce fusion reactions just before the bubbles broke up. As with cold fusion, most attempts to replicate their results have failed.

Conditions for Practical Fusion Yield

Two conditions must be met to achieve practical energy yields from fusion. First, the plasma temperature must be high enough that fusion reactions occur at a sufficient rate. Second, the plasma must be confined so that the energy released by fusion reactions, when deposited in the plasma, maintains its temperature against loss of energy by such phenomena as conduction, convection, and radiation. When these conditions are achieved, the plasma is said to be ignited. In the case of stars, or some approaches to fusion by magnetic confinement, a steady state can be achieved, and no energy beyond what is supplied from fusion reactions is needed to sustain the system. In other cases, such as the ICF approach, there is a large temperature excursion once fuel ignition is achieved. The energy yield can far exceed the energy required to attain plasma ignition conditions, but this energy is released in a burst, and the process has to be repeated roughly once every second for practical power to be produced.

The conditions for plasma ignition are readily derived. When fusion reactions occur in a plasma, the power released is proportional to the square of plasma ion density, n^2. The plasma loses energy when electrons scatter from positively charged ions, accelerating and radiating in the process. Such radiation is called bremsstrahlung and is proportional to $n^2 T^{1/2}$, where T is the plasma temperature. Other mechanisms by which heat can escape the plasma lead to a characteristic energy-loss time denoted by τ. The energy content of the plasma at temperature T is $3nkT$, where k is the Boltzmann constant. The rate of energy loss by mechanisms other than bremsstrahlung is thus simply $3nkT/\tau$. The energy balance of the plasma is the balance between the fusion energy heating the plasma and the energy-loss rate, which is the sum of $3nkT/\tau$ and the bremsstrahlung. The condition satisfying this balance is called the ignition condition. An equation relates the product of density and energy confinement time, denoted $n\tau$, to a function that depends only on the plasma temperature and the type of fusion reaction. For example, when the plasma is composed of deuterium and tritium, the smallest value of $n\tau$ required to achieve ignition is about 2×10^{20} particles per cubic metre times seconds, and the required temperature corresponds to an energy of about 25,000 eV. If the only energy losses are due to bremsstrahlung escaping from the plasma (meaning τ is infinite), the ignition temperature decreases to an energy level of 4,400 eV. Hence, the keys to generating usable amounts of fusion energy are to attain a sufficient plasma temperature and a sufficient confinement quality, as measured by the product $n\tau$. At a temperature equivalent to 10,000 eV, the $n\tau$ product must be about 3×10^{20} particles per cubic metre times seconds.

Magnetic fusion energy generally creates plasmas with a density of about 3×10^{20} particles per cubic metre, which is about 10^{-8} of normal density. Hence, the characteristic time for heat to escape must be greater than about one second. This is a measure of the required degree of magnetic insulation for the heat content. Under these conditions the plasma remains in energy balance and can operate continuously if the ash of the nuclear fusion, namely helium, is removed (otherwise it will quench the plasma) and fuel is replenished.

ICF creates plasmas of much higher density, generally between 10^{31} and 10^{32} particles per cubic metre, or 1,000 to 10,000 times the normal density. As such, the confinement time, or minimum burn time, can be as short as 20×10^{-12} second. The objective in ICF is to achieve a temperature equivalent of 4,400 eV at the centre of the highly compressed fuel mass, while still having sufficient mass left around the centre so that the disassembly time will excee the minimum burn time.

Nuclear Fission

Nuclear fission is the subdivision of a heavy atomic nucleus, such as that of uranium or plutonium, into two fragments of roughly equal mass. The process is accompanied by the release of a large amount of energy.

In nuclear fission the nucleus of an atom breaks up into two lighter nuclei. The process may take place spontaneously in some cases or may be induced by the excitation of the nucleus with a variety of particles (e.g., neutrons, protons, deuterons, or alpha particles) or with electromagnetic radiation in the form of gamma rays. In the fission process, a large quantity of energy is released, radioactive products are formed, and several neutrons are emitted. These neutrons can induce fission in a nearby nucleus of fissionable material and release more neutrons that can repeat the sequence, causing a chain reaction in which a large number of nuclei undergo fission and an enormous amount of energy is released. If controlled in a nuclear reactor, such a chain reaction can provide power for society's benefit. If uncontrolled, as in the case of the so-called atomic bomb, it can lead to an explosion of awesome destructive force.

The discovery of nuclear fission has opened a new era—the "Atomic Age." The potential of nuclear fission for good or evil and the risk/benefit ratio of its applications have not only provided the basis of many sociological, political, economic, and scientific advances but grave concerns as well. Even from a purely scientific perspective, the process of nuclear fission has given rise to many puzzles and complexities, and a complete theoretical explanation is still not at hand.

Fundamentals of the Fission Process

Structure and Stability of Nuclear Matter

The fission process may be best understood through a consideration of the structure and stability of nuclear matter. Nuclei consist of nucleons (neutrons and protons), the total number of which is equal to the mass number of the nucleus. The actual mass of a nucleus is always less than the sum of the masses of the free neutrons and protons that constitute it, the difference being the mass equivalent of the energy of formation of the nucleus from its constituents. The conversion of mass to energy follows Einstein's equation, $E = mc_2$, where E is the energy equivalent of a mass, m, and c is the velocity of light. This difference is known as the mass defect and is a measure of the total binding energy (and, hence, the stability) of the nucleus. This binding energy is released during the formation of a nucleus from its constituent nucleons and would have to be supplied to the nucleus to decompose it into its individual nucleon components.

A curve illustrating the average binding energy per nucleon as a function of the nuclear mass number is shown in figure. The largest binding energy (highest stability) occurs near mass number 56—the mass region of the element iron. Figure indicates that any nucleus heavier than mass number 56 would become a more stable system by breaking into lighter nuclei of higher binding energy, the difference in binding energy being released in the process. It should be noted that nuclei lighter than mass number 56 can gain in stability by fusing to produce a heavier nucleus of greater mass defect—again, with the release of the energy equivalent of the mass difference. It is the fusion of the lightest nuclei that provides the energy released by the Sun and constitutes the basis of the hydrogen, or fusion, bomb. Efforts to harness fusion reaction for power production have been actively pursued.

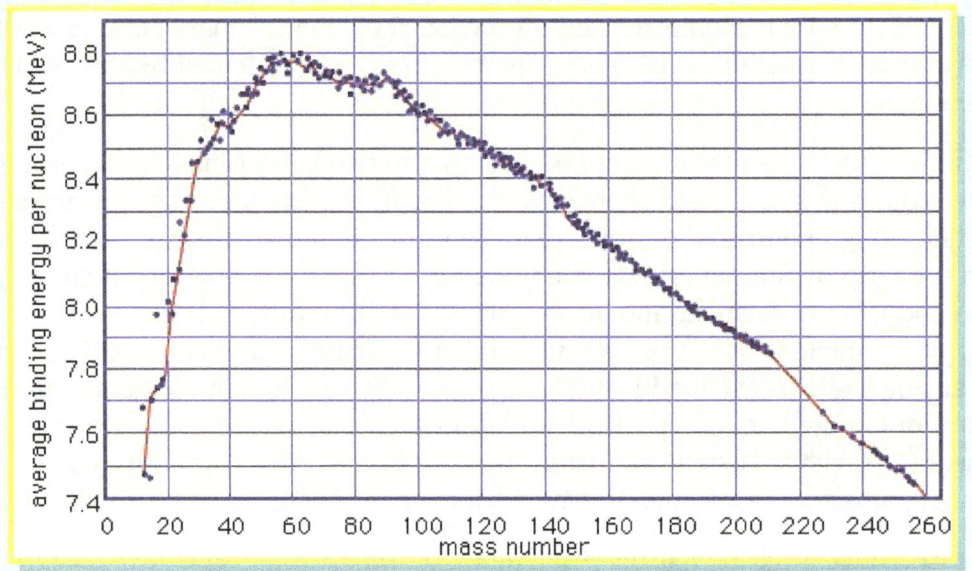

The average binding energy per nucleon as a function of
the mass number, A. The line connects the odd-A points.

On the basis of energy considerations alone, figure would indicate that all matter should seek its most stable configuration, becoming nuclei of mass number near 56. However, this does not happen, because barriers to such a spontaneous conversion are provided by other factors. A good qualitative understanding of the nucleus is achieved by treating it as analogous to a uniformly charged liquid drop. The strong attractive nuclear force between pairs of nucleons is of short range and acts only between the closest neighbours. Since nucleons near the surface of the drop have fewer close neighbours than those in the interior, a surface tension is developed, and the nuclear drop assumes a spherical shape in order to minimize this surface energy. (The smallest surface area enclosing a given volume is provided by a sphere.) The protons in the nucleus exert a long-range repulsive (Coulomb) force on each other because of their positive charge. As the number of nucleons in a nucleus increases beyond about 40, the number of protons must be diluted with an excess of neutrons to maintain relative stability.

If the nucleus is excited by some stimulus and begins to oscillate (i.e., deform from its spherical shape), the surface forces will increase and tend to restore it to a sphere, where the surface tension is at a minimum. On the other hand, the Coulomb repulsion decreases as the drop deforms and the protons are positioned farther apart. These opposing tendencies set up a barrier in the potential energy of the system, as indicated in figure.

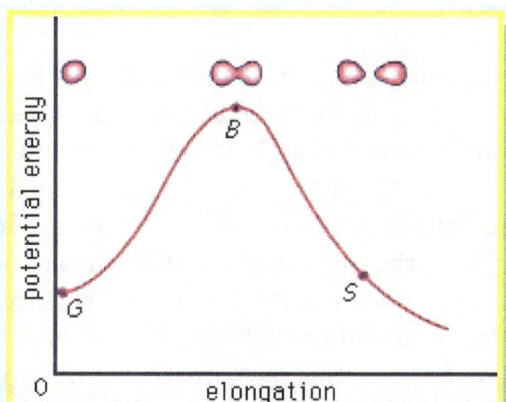

The potential energy as a function of elongation of a fissioning nucleus. G is the ground state of the nucleus; B is the top of the barrier to fission (called the saddle point); and S is the scission point. The nuclear shape at these points is shown at the top.

The curve in figure rises initially with elongation, since the strong, short-range nuclear force that gives rise to the surface tension increases. The Coulomb repulsion between protons decreases faster with elongation than the surface tension increases, and the two are in balance at point B, which represents the height of the barrier to fission. (This point is called the "saddle point" because, in a three-dimensional view of the potential energy surface, the shape of the pass over the barrier resembles a saddle.) Beyond point B, the Coulomb repulsion between the protons drives the nucleus into further elongation until at some point, S (the scission point), the nucleus breaks in two. Qualitatively, at least, the fission process is thus seen to be a consequence of the Coulomb repulsion between protons.

Induced Fission

The height and shape of the fission barrier are dependent on the particular nucleus being considered. Fission can be induced by exciting the nucleus to an energy equal to or greater than that of the barrier. This can be done by gamma-ray excitation (photofission) or through excitation of the nucleus by the capture of a neutron, proton, or other particle (particle-induced fission). The binding energy of a particular nucleon to a nucleus will depend on—in addition to the factors considered above—the odd–even character of the nucleus. Thus, if a neutron is added to a nucleus having an odd number of neutrons, an even number of neutrons will result, and the binding energy will be greater than for the addition of a neutron that makes the total number of neutrons odd. This "pairing energy" accounts in part for the difference in behaviour of nuclides in which fission can be induced with slow (low-energy) neutrons and those that require fast (higher-energy) neutrons. Although the heavy elements are unstable with respect to fission, the reaction takes place to an appreciable extent only if sufficient energy of activation is available to surmount the fission barrier. Most nuclei that are fissionable with slow neutrons contain an odd number of neutrons (e.g., uranium-233, uranium-235, or plutonium-239), whereas most of those requiring fast neutrons (e.g., thorium-232 or uranium-238) have an even number. The addition of a neutron in the former case liberates sufficient binding energy to induce fission. In the latter case, the binding energy is less and may be insufficient to surmount the barrier and induce fission. Additional energy must then be supplied in the form of the kinetic energy of the incident neutron. (In the case of thorium-232 or uranium-238, a neutron having about 1 MeV of kinetic energy is required).

Spontaneous Fission

The laws of quantum mechanics deal with the probability of a system such as a nucleus or an atom being in any of its possible states or configurations at any given time. A fissionable system (uranium-238, for example) in its ground state (i.e., at its lowest excitation energy and with an elongation small enough that it is confined inside the fission barrier) has a small but finite probability of being in the energetically favoured configuration of two fission fragments. In effect, when this occurs, the system has penetrated the barrier by the process of quantum mechanical tunneling. This process is called spontaneous fission because it does not involve any outside influences. In the case of uranium-238, the process has a very low probability, requiring more than 10^{15} years for half of the material to be transformed (its so-called half-life) by this reaction. On the other hand, the probability for spontaneous fission increases dramatically for the heaviest nuclides known and becomes the dominant mode of decay for some—those having half-lives of only fractions of a second. In fact, spontaneous fission becomes the limiting factor that may prevent the formation of still heavier (super-heavy) nuclei.

A pictorial representation of the sequence of events in the fission of a heavy nucleus is given in figure. The approximate time elapse between stages of the process is indicated at the bottom of the figure.

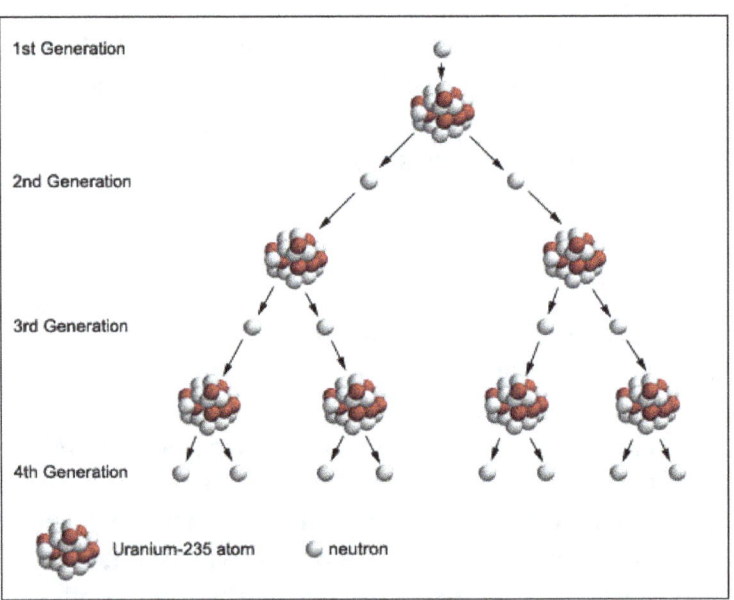

Sequence of events in the fission of a uranium nucleus by a neutron.

The Phenomenology of Fission

When a heavy nucleus undergoes fission, a variety of fragment pairs may be formed, depending on the distribution of neutrons and protons between the fragments. This leads to probability distribution of both mass and nuclear charge for the fragments. The probability of formation of a particular fragment is called its fission yield and is expressed as the percentage of fissions leading to it.

The separated fragments experience a large Coulomb repulsion due to their nuclear charges, and they recoil from each other with kinetic energies determined by the fragment charges and the distance between the charge centres at the time of scission. Variations in these parameters lead to a distribution of kinetic energies, even for the same mass split.

The initial velocities of the recoiling fragments are too fast for the outer (atomic) electrons of the fissioning atom to keep pace, and many of them are stripped away. Thus, the nuclear charge of the fragment is not fully neutralized by the atomic electrons, and the fission fragments fly apart as highly charged atoms. As the nucleus of the fragment adjusts from its deformed shape to a more stable configuration, the deformation energy (i.e., the energy required to deform it) is recovered and converted into internal excitation energy, and neutrons and prompt gamma rays (an energetic form of electromagnetic radiation given off nearly coincident with the fission event) may be evaporated from the moving fragment. The fast-moving, highly charged atom collides with the atoms of the medium through which it is moving, and its kinetic energy is transferred to ionization and heating of the medium as it slows down and comes to rest. The range of fission fragments in air is only a few centimetres.

During the slowing-down process, the charged atom picks up electrons from the medium and becomes neutral by the time it stops. At this stage in the sequence of events, the atom produced is called a fission product to distinguish it from the initial fission fragment formed at scission. Since a few neutrons may have been lost in the transition from fission fragment to fission product, the two may not have the same mass number. The fission product is still not a stable species but is radioactive, and it finally reaches stability by undergoing a series of beta decays, which may vary over a time scale of fractions of a second to many years. The beta emission consists of electrons and antineutrinos, often accompanied by gamma rays and X-rays.

The distributions in mass, charge, and kinetic energy of the fragments have been found to be dependent on the fissioning species as well as on the excitation energy at which the fission act occurs. Many other aspects of fission have been observed, adding to the extensive phenomenology of the process and providing an intriguing set of problems for interpretation. These include the systematics of fission cross sections (a measure of the probability for fission to occur); the variation of the number of prompt neutrons emitted as a function of the fissioning species and the particular fragment mass split; the angular distribution of the fragments with respect to the direction of the beam of particles inducing fission; the systematics of spontaneous fission half-lives; the occurrence of spontaneous fission isomers (excited states of the nucleus); the emission of light particles (hydrogen-3, helium-3, helium-4, etc.) in small but significant numbers in some fission events; the presence of delayed neutron emitters among the fission products; the time scale on which the various stages of the process take place; and the distribution of the energy release in fission among the particles and radiations produced.

Fission Fragment Mass Distributions

The distribution of the fragment masses formed in fission is one of the most striking features of the process. It is dependent on the mass of the fissioning nucleus and the excitation energy at which the fission occurs. At low excitation energy, the fission of such nuclides as uranium-235 or plutonium-239 is asymmetric; i.e., the fragments are formed in a two-humped probability (or yield) distribution favouring an unequal division in mass. This is illustrated in figure, the light group of fragment masses shifts to higher mass numbers as the mass of the fissioning nucleus increases, whereas the position of the heavy group remains nearly stationary. As the excitation energy of the fission increases, the probability for a symmetric mass split increases, while that for asymmetric division decreases. Thus, the valley between the two peaks increases in probability (yield of formation), and at high excitations the mass distribution becomes single-humped, with the maximum yield at symmetry. Radium isotopes show interesting triple-humped mass distributions, and nuclides lighter than radium show a single-humped, symmetric mass distribution. (These nuclides, however, require a relatively high activation energy to undergo fission.) For very heavy nuclei in the region of fermium-260, the mass-yield curve becomes symmetric (single-humped) even for spontaneous fission, and the kinetic energies of the fragments are unusually high. An understanding of these mass distributions has been one of the major puzzles of fission, and a complete theoretical interpretation is still lacking, albeit much progress has been made.

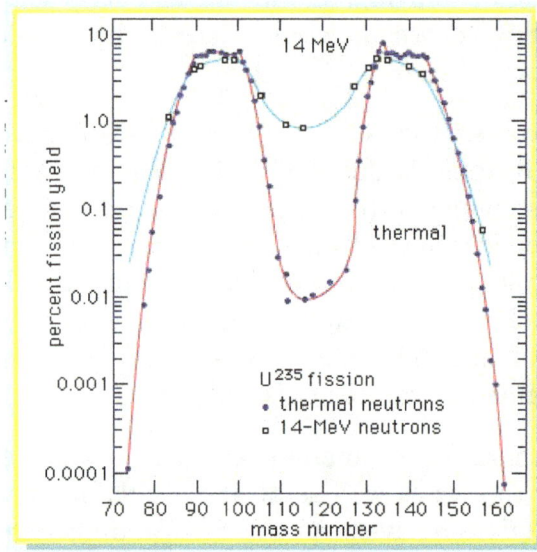

Mass distribution dependence on the energy excitation in the fission of uranium-235. At still higher energies, the curve becomes single-humped, with a maximum yield for symmetric mass splits.

Fission Decay Chains and Charge Distribution

In order to maintain stability, the neutron-to-proton (n/p) ratio in nuclei must increase with increasing proton number. The ratio remains at unity up to the element calcium, with 20 protons. It then gradually increases until it reaches a value of about 1.5 for the heaviest elements. When a heavy nucleus fissions, a few neutrons are emitted; however, this still leaves too high an n/p ratio in the fission fragments to be consistent with stability for them. They undergo radioactive decay and reach stability by successive conversions of neutrons to protons with the emission of a negative electron (called a beta particle, β-) and an antineutrino. The mass number of the nucleus remains the same, but the nuclear charge (atomic number) increases by one, and a new element is formed for each such conversion. The successive beta decays constitute an isobaric, fission-product decay chain for each mass number. The half-lives for the decay of the radioactive species generally increase as they approach the stable isobar of the chain. (Species of the same element characterized by the same nuclear charge, Z [number of protons], but differing in their number of neutrons [and therefore in mass number A] are called isotopes. Species that have the same mass number, A, but differ in Z are known as isobars.)

For a typical mass split in the neutron-induced fission of uranium-235, the complementary fission-product masses of 93 and 141 may be formed following the emission of two neutrons from the initial fragments. The division of charge (i.e., protons) between the fragments represents an important parameter in the fission process. Thus, for the mass numbers 93 and 141, the following isobaric fission-product decay chains would be formed (the half-lives for the beta-decay processes are indicated above the arrows):

$$^{93}_{36}\text{Kr} \xrightarrow[\beta^-]{1.3 \text{ sec}} {}^{93}_{37}\text{Rb} \xrightarrow[\beta^-]{6 \text{ sec}} {}^{93}_{38}\text{Sr} \xrightarrow[\beta^-]{7.5 \text{ min}} {}^{93}_{39}\text{Y} \xrightarrow[\beta^-]{10 \text{ hr}} {}^{93}_{40}\text{Zr} \xrightarrow[\beta^-]{1.5 \times 10^6 \text{ yr}} {}^{93m}_{41}\text{Nb}$$

$$\xrightarrow{\beta^-} {}^{93}_{41}\text{Nb (stable)} \quad \downarrow 13 \text{ yr}$$

$$^{141}_{54}\text{Xe} \xrightarrow[\beta^-]{1.7 \text{ sec}} {}^{141}_{55}\text{Cs} \xrightarrow[\beta^-]{25 \text{ sec}} {}^{141}_{56}\text{Ba} \xrightarrow[\beta^-]{18 \text{ min}} {}^{141}_{57}\text{La} \xrightarrow[\beta^-]{3.9 \text{ hr}} {}^{141}_{58}\text{Ce} \xrightarrow[\beta^-]{32.5 \text{ d}} {}^{141}_{59}\text{Pr (stable)}$$

(The left subscript on the element symbol denotes Z, while the superscript denotes A.) The 92 protons of the uranium nucleus must be conserved, and complementary fission-product pairs—such as krypton-36 with barium-56, rubidium-37 with cesium-55, or strontium-38 with xenon-54—would be possible.

The percentage of fissions in which a particular nuclide is formed as a primary fission product (i.e., as the direct descendant of an initial fragment following its de-excitation) is called the independent yield of that product. The total yield for any nuclide in the isobaric decay chain is the sum of its independent yield and the independent yields of all of its precursors in the chain. The total yield for the entire chain is called the cumulative yield for that mass number.

Extensive radiochemical investigations have suggested that the most probable charge division is one that is displaced from stability about the same distance in both chains. This empirical observation is called the equal charge displacement (ECD) hypothesis, and it has been confirmed by several physical measurements. In the above example the ECD would predict the most probable charges at about rubidium-37 and cesium-55. A strong shell effect modifies the ECD expectations for fragments having 50 protons. The dispersion of the charge formation probability about the most probable charge (Zp) is rather narrow and approximately Gaussian in shape and is nearly independent of the mass split as well as of the fissioning species. The most probable charge for an isobaric chain is a useful concept in the description of the charge dispersion, and it need not have an integral value. As the energy of fission increases, the charge division tends toward maintaining the n/p ratio in the fragments the same as that in the fissioning nucleus. This is referred to as an unchanged charge distribution.

Prompt Neutrons in Fission

The average number of neutrons emitted per fission (represented by the symbol $\bar{v}$) varies with the fissioning nucleus. It is about 2.0 for the spontaneous fission of uranium-238 and 4.0 for that of fermium-257. In the thermal-neutron induced fission of uranium-235, $\bar{v} = 2.4$. The actual number of neutrons emitted, however, varies with each fission event, depending on the mass split. Although there is still controversy regarding the number of neutrons emitted at the instant of scission, it is generally agreed that most of the neutrons are given off by the recoiling fission fragments soon after scission occurs. The number of neutrons emitted from each fragment depends on the amount of energy the fragment possesses. The energy can be in the form of internal excitation (heat) energy or stored as energy of deformation of the fragment to be released when the fragment returns to its stable equilibrium shape.

Delayed Neutrons in Fission

A few of the fission products have beta-decay energies that exceed the binding energy of a neutron in the daughter nucleus. This is likely to happen when the daughter nucleus contains one or two neutrons more than a closed shell of 50 or 82 neutrons, since these "extra" neutrons are more loosely bound. The beta decay of the precursor may take place to an excited state of the daughter from which a neutron is emitted. The neutron emission is "delayed" by the beta-decay half-life of the precursor. Six such delayed neutron emitters have been identified, with half-lives varying from about 0.5 to 56 seconds. The yield of the delayed neutrons is only about 1 percent of that of the prompt neutrons, but they are very important for the control of the chain reaction in a nuclear reactor.

Energy Release in Fission

The total energy release in a fission event may be calculated from the difference in the rest masses of the reactants (e.g., ^{235}U + n) and the final stable products (e.g., ^{93}Nb + ^{141}Pr + 2n). The energy equivalent of this mass difference is given by the Einstein relation, $E = mc^2$. The total energy release depends on the mass split, but a typical fission event would have the total energy release distributed approximately as follows for the major components in the thermal neutron-induced fission of uranium-235:

Energy component	Number per fission	Total energy
Kinetic energy of fission fragments	2	170 MeV
Kinetic energy of prompt neutrons	2.5	5
Binding energy from capture of prompt neutron	2.5	12
Prompt gamma rays	8	8
Total = 195		

(The energy release from the capture of the prompt neutrons depends on how they are finally stopped, and some will escape the core of a nuclear reactor.)

This energy is released on a time scale of about 10^{-12} second and is called the prompt energy release. It is largely converted to heat within an operating reactor and is used for power generation. Also, there is a delayed release of energy from the radioactive decay of the fission products varying in half-life from fractions of a second to many years. The shorter-lived species decay in the reactor, and their energy adds to the heat generated; however, the longer-lived species remain radioactive and pose a problem in the handling and disposition of the reactor fuel elements when they need to be replaced. The antineutrinos that accompany the beta decay of the fission products are unreactive, and their kinetic energy (about 10 MeV per fission) is not recovered. Overall, about 200 MeV of energy per fission may be recovered for power applications.

Fission Theory

Nuclear fission is a complex process that involves the rearrangement of hundreds of nucleons in a single nucleus to produce two separate nuclei. A complete theoretical understanding of this reaction would require a detailed knowledge of the forces involved in the motion of each of the nucleons through the process. Since such knowledge is still not available, it is necessary to construct simplified models of the actual system to simulate its behaviour and gain as accurate a description as possible of the steps in the process. The successes and failures of the models in accounting for the various observations of the fission process can provide new insights into the fundamental physics governing the behaviour of real nuclei, particularly at the large nuclear deformations encountered in a nucleus undergoing fission.

The framework for understanding nuclear reactions is analogous to that for chemical reactions and involves the concept of a potential-energy surface on which the reaction occurs. The driving force for physical or chemical reactions is the tendency to lower the potential energy and increase the stability of the system. Thus, for example, a stone at the top of a hill will roll down the hill, converting its potential energy at the top to kinetic energy of motion, and will come to rest at the bottom in a more stable state of lower potential energy. The potential energy is calculated as a function of various parameters of the system being studied. In the case of fission, the potential energy may be calculated as a function of the shape of the system as it proceeds over the barrier to the scission point, and the path of lowest potential energy may be determined.

An exact calculation of the nuclear potential energy is not yet possible, and it is to approximate this calculation that various models have been constructed to simulate the real system. Some of the models were developed to address aspects of nuclear structure and spectroscopy as well as features of nuclear reactions, and they also have been employed in attempts to understand the complexity of nuclear fission. The models are based on different assumptions and approximations of the nature of the nuclear forces and the dynamics of the path to scission. No one model can account for all of the extensive phenomenology of fission, but each addresses different aspects of the process and provides a foundation for further development toward a complete theory.

Nuclear Models and Nuclear Fission

The nucleus exhibits some properties that reflect the collective motion of all its constituent nucleons as a unit, as well as other properties that are dependent on the motion and state of the individual nucleons.

The analogy of the nucleus to a drop of an incompressible liquid was first suggested by George Gamow in 1935 and later adapted to a description of nuclear reactions (by Niels Bohr; and Bohr and Fritz Kalckar) and to fission (Bohr and John A. Wheeler; and Yakov Frenkel). Bohr proposed the so-called compound nucleus description of nuclear reactions, in which the excitation energy of the system formed by the absorption of a neutron or photon, for example, is distributed among a large number of degrees of freedom of the system. This excited state persists for a long time relative to the periods of motion of nucleons across the nucleus and then decays by emission of radiation, the evaporation of neutrons or other particles, or by fission. The liquid-drop model of the nucleus accounts quite well for the general collective behaviour of nuclei and provides an understanding of the fission process on the basis of the competition between the cohesive nuclear force and the disruptive Coulomb repulsion between protons. It predicts, however, a symmetric division of mass in fission, whereas an asymmetric mass division is observed. Moreover, it does not provide an accurate description of fission barrier systematics or of the ground-state masses of nuclei. The liquid-drop model is particularly useful in describing the behaviour of highly excited nuclei, but it does not provide an accurate description for nuclei in their ground or low-lying excited states. Many versions of the liquid-drop model employing improved sets of parameters have been developed. However, investigators have found that mass asymmetry and certain other features in fission cannot be adequately described on the basis of the collective behaviour posited by such models alone.

A preference for the formation of unequal masses (i.e., an asymmetric division) was observed early in fission research, and it has remained the most puzzling feature of the process to account for. Investigators have invoked various models other than that of the liquid drop in an attempt to address this question. Dealing with the mutual interaction of all the nucleons in a nucleus has been simplified by treating it as if it were equivalent to the interaction of one particle with an average spherical static potential field that is generated by all the other nucleons. The methods of quantum mechanics provide the solution for the motion of a nucleon in such a potential. A characteristic set of energy levels for neutrons and protons is obtained, and, analogous to the set of levels of the electrons in an atom, the levels group themselves into shells at certain so-called magic numbers of nucleons. (For both neutrons and protons, these numbers are 2, 8, 20, 28, 50, 82, and 126.) Shell closures at these nuclear numbers are marked by especially strong binding, or extra stability. This constitutes the essence of the spherical-shell model (sometimes called the independent-particle, or single-particle, model), as developed by Maria Goeppert Mayer and J. Hans D. Jensen and their colleagues. It accounts well for ground-state masses and spins and for the existence of isomeric nuclear states (excited states having measurable half-lives) that occur when nuclear levels of widely differing spins lie relatively close to each other. The agreement with observations is excellent for spherical nuclei with nucleon numbers near the magic shell numbers. The spherical-shell model, however, does not agree well with the properties of nuclei that have other nucleon numbers—e.g., the nuclei of the lanthanide and actinide elements, with nucleon numbers between the magic numbers.

In the lanthanide and actinide nuclei, the ground state is not spherical but rather deformed into a prolate spheroidal shape—that of a football or watermelon. For such nuclei, the allowed states of motion of a nucleon must be calculated in a potential having a symmetry corresponding to a spheroid rather than a sphere. This was first done by Aage Bohr, Ben R. Mottelson, and Sven G. Nilsson in 1955, and the level structure was calculated as a function of the deformation of the nucleus. A spheroid has three axes of symmetry, and it can rotate in space as a unit about any one of them. The rotation can occur independent of the internal state of excitation of the individual nucleons. Various modes of vibration of the spheroid also may take place. Since this deformed shell model has components of both the independent-particle motion and the collective motion of the nucleus as a whole (i.e., rotations and vibrations), it is sometimes referred to as the unified model.

In Aage Bohr's application of the unified model to the fission process, the sequence of potential-energy surfaces for the excited states of the system are considered to be functions of a deformation parameter (i.e., elongation) characterizing the motion toward fission and evaluated at the saddle point. As the system passes over the saddle point, most of its excitation energy is used up in deforming the nucleus, and the system remains "cold"; i.e., it manifests little excitation, or heat, energy. Thus, only the low-lying excited states are available to the system. The spin and parity of the particular state (or channel) in which the system exists as it passes over the saddle point are then expected to determine the fission properties. In this channel (or transition-state) analysis of fission, a number of characteristics of the process are qualitatively accounted for. Hence, fission thresholds would depend on the spin and parity of the compound nuclear state, the fission fragment angular distribution would be governed by the collective rotational angular momentum of the state, and asymmetry in the mass distribution would result from passage over the barrier in a state of negative parity (which does not possess reflection symmetry). This model gives a good qualitative interpretation of many fission phenomena, but it must assume that at least some of the properties of the transition state at the saddle point are not altered by dynamical considerations in the descent of the system to the scission point. It is the only model that provides a satisfactory interpretation of the angular distributions of fission fragments, and it has attractive features that must be included in any complete theory of fission.

The first application of the spherical-shell model to fission was the recognition that the positions of the peaks in the fission mass distribution correlated fairly well with the magic numbers and suggested a qualitative interpretation of the asymmetric mass division. Thus, a preference for the formation of nuclei with neutron numbers close to 82 would favour the formation of nuclides near the peak in the heavy group and would thus determine the mass split for the fissioning system. Some extra stability for nuclear configurations of 50 protons would also be expected, but this is not particularly evident. In fact, the so-called doubly magic nucleus tin-132, with 50 protons and 82 neutrons, has a rather low yield in low-energy fission.

A more quantitative application of the spherical-shell model to fission was undertaken by Peter Fong in the United States in 1956. He related the probability of formation of a given pair of fragments to the available density of states for that pair of fragments at the scission point in a statistical-model approach. A model of this sort predicts that the system, in its random motions, will experience all possible configurations and so will have a greater probability of being in the region where the greatest number of such configurations (or states) is concentrated. The model assumes that the potential energy at the saddle point is essentially all converted to excitation energy and that a statistical equilibrium among all possible states is established at the scission point. The extra binding energy for closed-shell nuclei leads to a higher density of states at a given excitation energy than is present for other nuclei and, hence, leads to a higher probability of formation. An asymmetric mass distribution in good agreement with that observed for the neutron-induced fission of uranium-235 is obtained. Moreover, the changes in the mass distribution with an increased excitation energy of fission (e.g., an increase in the probability of symmetric fission relative to asymmetric fission) are accounted for by the decrease in importance of the shell effects as the excitation energy increases. Other features of the fission process also are qualitatively explained; however, extensive changes in the parameters of the model are required to obtain agreement with experiments for other fissionable nuclides. Then, too, there are fundamental problems concerning the validity of some of the basic assumptions of the model.

The fundamental question as to the validity of models that evaluate the properties of the system at the scission point (the so-called scission-point models of fission) is whether the system remains long enough at this point on the steep decline of the potential-energy surface for a quasi-equilibrium condition to be established. There is some evidence that such a condition may indeed prevail, but it is not clearly established. Nonetheless, such models have proved quite useful in interpreting observations of mass, charge, and kinetic energy distributions, as well as of neutron emission dependence on fragment mass. It seems very likely that the fragment shell structure plays a significant role in determining the course of the fission process.

Although the single-particle models provide a good description of various aspects of nuclear structure, they are not successful in accounting for the energy of deformation of nuclei (i.e., surface energy), particularly at the large deformations encountered in the fission process. A major breakthrough occurred when a hybrid model incorporating shell effects as a correction to the potential energy of the liquid-drop model was proposed by the Russian physicist V.M. Strutinskii in 1967. This approach retains the dominant collective surface and Coulomb effects while adding shell and pairing corrections that depend on deformation. Shell corrections of several million electron volts are calculated, and these can have a significant effect on a liquid-drop barrier of about 5 MeV. The nucleon numbers at which the shells appear depend on the deformation and may differ from the spherical model magic numbers. In the vicinity of the fission barrier, the shells introduce structure in the liquid-drop potential-energy curve, as illustrated in figure. The relative heights and widths of the two peaks vary with the mass and charge of the fissioning system.

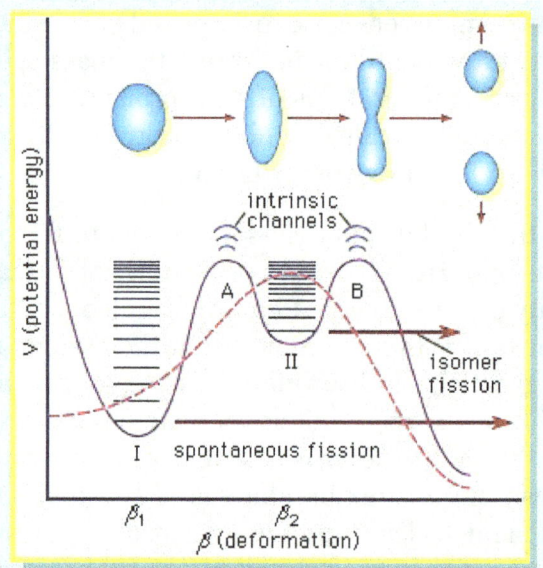

Schematic illustrations of single-humped and double-humped fission barriers. The former are represented by the dashed line and the latter by the continuous line. Intrinsic excitations in the first and second wells at deformations β_1 and β_2 are designated class I and class II states, respectively. Intrinsic channels at the two barriers also are illustrated. The transition in the shape of the nucleus as a function of deformation is schematically represented in the upper part of the figure. Spontaneous fission of the ground state and isomeric state occurs from the lowest energy class I and class II states, respectively.

The double-humped barrier provides a satisfactory explanation for a number of puzzling observations in fission. The existence of short-lived, spontaneous fission isomers, for example, is understood as the consequence of the population of states in the second well (class II). These isomers have a much smaller barrier to penetrate and so exhibit a much shorter spontaneous fission half-life. The change in shape associated with these states, as compared to class I states, also hinders a rapid return to the ground state by gamma emission. (Class II states are also called shape isomers.) The systematics of neutron-induced fission cross sections and structure in some fission-fragment angular distributions also find an interpretation in the implications of the double-humped barrier.

The Strutinskii procedure provided a strong stimulus for calculations of the potential-energy surfaces appropriate to fissioning systems, since it provided a consistent and useful prescription for treating both the macroscopic (liquid-drop) and microscopic (single-particle) effects in deformed nuclei. Many calculations of the potential-energy surface employing different model potentials and parameters have been carried out as functions of the shapes of the system. The work of the American nuclear physicists W.J. Swiatecki, James R. Nix, and their collaborators has been particularly noteworthy in such studies, which also include some attempts to treat the dynamical evolution of the fission process.

Calculations for the actinide elements indicate that, at deformations corresponding to the second barrier, the potential energy for asymmetric mass splits is lower than that for symmetric ones; hence, the former are favoured at that stage of the process. For larger deformations, however, a single potential does not represent the incipient formation of two fragments very well. In fact, a discontinuity occurs at the scission point, and the results of the calculation depend on whether the scission configuration is treated as one nucleus or as two separate nuclei.

A two-centre potential may also be used to represent the nature of the forces at work in a fissioning nucleus. In such a model, the potential energy surfaces are represented by two overlapping spheres or spheroids. It is equivalent to a one-centre potential when there is a complete overlap at small deformations, and it has the correct asymptotic behaviour as the nascent fragments separate. This approach indicates a preformation of the final shell structure of the fragments early in the process.

Although the validity of the assumptions inherent in scission-point models may be in question, the results obtained with them are in excellent agreement with observation. Representative of such a model is the Argonne Scission-Point model, which uses a macroscopic-microscopic calculation with deformed fragment shell and pairing corrections to determine the potential energy of a system of two nearly touching spheroids and which includes their interaction in terms of a neck connecting them. Models of this kind provide a simple approach to a highly quantitative and detailed study of the dependence of the probability of formation of a given fragment pair on the neutron and proton number and on the deformation in each fragment. They account very well for the mass, charge, and kinetic-energy distributions and the neutron-emission dependence on mass number for a broad range of fissioning nuclei. The scission-point models, however, do not address questions of fission probability or the angular distributions of the fragments. As the fission-excitation energy increases, the shell correction diminishes and the macroscopic (liquid-drop) behaviour dominates.

Nuclides in the region of fermium-264 have been observed to undergo symmetric fission with unusually high fragment kinetic energies. This appears to be the consequence of the stability for the magic number configurations of 50 protons and 82 neutrons. The formation of two doubly magic fragments of tin-132 is strongly favoured energetically, whereas the formation of only one such fragment in the low-energy fission of uranium or plutonium isotopes is not. The fragments of tin-132 are spherical rather than deformed, and a more compact configuration at the scission point (with the charge centres closer together) leads to higher fragment kinetic energies.

It is evident that shell effects, both in the fissioning system at the saddle point and in the deformed fragments near the scission point, are important in interpreting many of the features of the fission process. The stage of the process at which the various fragment distributions are determined is, however, not clearly established. All the components of a reasonable understanding of fission seem to be at hand, but they have yet to be synthesized into a complete, dynamic theory.

Considerations of the dynamics of the descent of the system on the potential-energy surface from the saddle point to the scission point involve two extreme points of view. An "adiabatic" approximation may be valid if the collective motion of the system is considered to be so slow—or the coupling between the collective and internal single-particle degrees of freedom (i.e., between macroscopic and microscopic behaviour) so weak—that the fast single-particle motions can readily adjust to the changes in shape of the fissioning nucleus as it progresses toward scission. In this case, the changes in the system take place without the gain or loss of heat energy. The decrease in potential energy between the saddle and scission points will then appear primarily in the collective degrees of freedom at scission and be associated with the kinetic energy of the relative motion of the nascent fragments (referred to as pre-scission kinetic energy). On the other hand, if the collective motion toward scission is relatively fast or the coupling-to-particle motion stronger, collective energy can be transformed into internal excitation (heat) energy of the nucleons. (This is analogous to heating in the motion of a viscous fluid.) In such a "non-adiabatic" process the mixing among the single-particle degrees of freedom may be sufficiently complete that a statistical model may be applicable at the scission point. Either extreme represents an approximation of complex behaviour, and some experimental evidence in support of either interpretation may be advanced. As in most such instances in nature, the truth probably lies somewhere between the extremes, with both playing some role in the fission process.

Fission Chain Reactions and their Control

The emission of several neutrons in the fission process leads to the possibility of a chain reaction if at least one of the fission neutrons induces fission in another fissile nucleus, which in turn fissions and emits neutrons to continue the chain. If more than one neutron is effective in inducing fission in other nuclei, the chain multiplies more rapidly. The condition for a chain reaction is usually expressed in terms of a multiplication factor, k, which is defined as the ratio of the number of fissions produced in one step (or neutron generation) in the chain to the number of fissions in the preceding generation. If k is less than unity, a chain reaction cannot be sustained. If k = 1, a steady-state chain reaction can be maintained; and if k is greater than 1, the number of fissions increases at each step, resulting in a divergent chain reaction. The term critical assembly is applied to a configuration of fissionable material for which k = 1; if k > 1, the assembly is said to be supercritical. A critical assembly might consist of the fissile material in the form of a metal or oxide, a moderator to slow the fission neutrons, and a reflector to scatter neutrons that would otherwise be lost back into the assembly core.

In a fission bomb it is desirable to have k as large as possible and the time between steps in the chain as short as possible so that many fissions occur and a large amount of energy is generated within a brief period ($\sim 10-7$ second) to produce a devastating explosion. If one kilogram of uranium-235 were to fission, the energy released would be equivalent to the explosion of 20,000 tons of the chemical explosive trinitrotoluene (TNT). In a controlled nuclear reactor, k is kept equal to unity for steady-state operation. A practical reactor, however, must be designed with k somewhat greater than unity. This permits power levels to be increased if desired, as well as allowing for the following: the gradual loss of fuel by the fission process; the buildup of "poisons" among the fission products being formed that absorb neutrons and lower the k value; and the use of some of the neutrons produced for research studies or the preparation of radioactive species for various applications. The value of k is controlled during the operation of a reactor by the positioning of movable rods made of a material that readily absorbs neutrons (i.e., one with a high neutron-capture cross section), such as boron, cadmium, or hafnium. The delayed-neutron emitters among the fission products increase the time between successive neutron generations in the chain reaction and make the control of the reaction easier to accomplish by the mechanical movement of the control rods.

Fission reactors can be classified by the energy of the neutrons that propagate the chain reaction. The most common type, called a thermal reactor, operates with thermal neutrons (those having the same energy distribution as gas molecules at ordinary room temperatures). In such a reactor, the fission neutrons produced (with an average kinetic energy of more than 1 MeV) must be slowed down to thermal energy by scattering from a moderator, usually consisting of ordinary water, heavy water (D_2O), or graphite. In another type, termed an intermediate reactor, the chain reaction is maintained by neutrons of intermediate energy, and a beryllium moderator may be used. In a fast reactor, fast fission neutrons maintain the chain reaction, and no moderator is needed. All of the reactor types require a coolant to remove the heat generated; water, a gas, or a liquid metal may be used for this purpose, depending on the design needs.

Uses of Fission Reactors and Fission Products

A nuclear reactor is essentially a furnace used to produce steam or hot gases that can provide heat directly or drive turbines to generate electricity. Nuclear reactors are employed for commercial electric-power generation throughout much of the world and as a power source for propelling submarines and certain kinds of surface vessels. Another important use for reactors is the utilization of their high neutron fluxes for studying the structure and properties of materials and for producing a broad range of radionuclides, which, along with a number of fission products, have found many different applications. Heat generated by radioactive decay can be converted into electricity through the thermoelectric effect in semiconductor materials and thereby produce what is termed an atomic battery. When powered by either a long-lived beta-emitting fission product (e.g., strontium-90 or promethium-147) or one that emits alpha particles (plutonium-238 or curium-244), these batteries are a particularly useful source of energy for cardiac pacemakers and for instruments employed in remote unmanned facilities, such as those in outer space, the polar regions of the Earth, or the open seas.

3

Nuclear Materials

Substances required to carry out a nuclear reaction are termed as nuclear materials. Some of the materials that are used in nuclear reaction are uranium, plutonium, thorium, and their isotopes. This chapter closely examines these nuclear materials to provide an extensive understanding of this subject.

Nuclear materials are the key ingredients in nuclear weapons. They include fissile, fussionable and source materials. Fissile materials are those which are composed of atoms that can be split by neutrons in a self-sustaining chain-reaction to release energy, and include plutonium-239 and uranium-235. Fussionable materials are those in which the atoms can be fused in order to release energy, and include deuterium and tritium. Source materials are those which are used to boost nuclear weapons by providing a source of additional atomic particles for fission. They include tritium, polonium, beryllium, lithium-6 and helium-3.

Plutonium

Plutonium is not found naturally in significant quantities. It is produced in a nuclear reactor through the absorption of neutrons by Uranium 238. The Plutonium emerges from a nuclear reactor as part of the mix in spent nuclear fuel, along with unused uranium and other highly radioactive fission products. To get plutonium into a usable form, a second key facility, a reprocessing plant, is needed to chemically separate out the plutonium from the other materials in spent fuel.

Once plutonium is separated, it can be processed and fashioned into the fission core of a nuclear weapon, called a "pit". Nuclear weapons typically require three to five kilograms of plutonium. Plutonium can also be converted into an oxide and mixed with uranium dioxide to form mixed-oxide (MOX) fuel for nuclear reactors. Britain, France, Russia, India, Japan, Israel and China operate reprocessing plants to obtain plutonium (the last two only for military purposes). U.S. plutonium production reactors were shut down in 1988.

A number of isotopes of plutonium are produced in a reactor, the most common being Pu-239 which is easily fissionable, and Pu-240 which is not. The relative proportion of Pu-239 determines the weapons grade of the plutonium. Reactor grade Pu, i.e. Pu with 18% or more Pu-240, can still be used to make a "crude" nuclear bomb.

Plutonium is an alpha particle emitter and so does not penetrate the skin. However, when ingested into the body, plutonium is incredibly toxic as alpha particles cause a very high rate cell damage. It is possible, for example, to contract lung cancer from one millionth of a gram.

Uranium

Uranium occurs naturally in underground deposits consisting of a mixture of 0.7% uranium-235, which is easily fissionable, and about 99.3% uranium-238, which is not fissionable. Nuclear weapons require "enrichment" to increase the proportion of U235 to 90% or more. This is called Highly Enriched Uranium (HEU). Nuclear reactors require enrichment to about 3 - 5 % of U-235. This is called Low Enriched Uranium (LEU).

HEU can be combined with plutonium to form the "pit", or core of a nuclear weapon, or it can be used alone as the nuclear explosive. The bomb dropped on Hiroshima used only HEU. About 15-20 kgs of HEU are sufficient to make a bomb without plutonium.

Tritium

Tritium is a relatively rare form of hydrogen isotope with an atomic mass of three (one proton and two neutrons). It is used commercially, but only in minute quantities, for medical diagnostics and sign illumination. Tritium's primary function is to boost the yield of both fission and thermonuclear weapons. It is produced in fission reactors and high-energy accelerators by bombarding lithium or lithium compounds with high energy neutrons. Tritium decays rapidly with a half-life of 12.5 years, and thus must be replenished over time. For example, the U.S. has produced 225 kilograms since 1955. This has now decayed to an inventory of 75 kilograms.

Deuterium

Deuterium is a stable, naturally-occurring isotope of hydrogen with an atomic mass of two (one proton and one neutron). There is approximately 1 part of deuterium to 5000 parts of normal hydrogen found in nature. Deuterium is sometimes called heavy hydrogen. In thermonuclear bombs deuterium is fused with tritium to release energy.

Insecure Nuclear Materials

Nuclear materials are easier to monitor than materials suitable for chemical and biological weapons. This is because the key materials - Pu239 and HEU - require complex facilities to isolate. Even so, there are some difficulties.

The International Atomic Energy Agency (IAEA) has established a regime of safeguards on nuclear facilities in order to prevent diversion of fissile material for weapons purposes. Non-nuclear weapon States (NNWS) parties to the Non-Proliferation Treaty are required to sign safeguards agreements with the IAEA. In 1991 the discovery of Iraq's nuclear weapons program indicated shortcomings in the safeguards system. The IAEA thus developed a strengthened safeguards system and invited NNWS to join. However, not all NNWS parties to the NPT have joined. More significantly, the non-parties to the NPT and the NWS are not required to place their facilities under IAEA safeguards. The possibility of States diverting nuclear materials for weapons purposes therefore continues to exist.

In addition, there are large stockpiles of fissile material, and the security of some of this material is under question. In August 1994 German police confiscated a suitcase used to smuggle plutonium from Moscow to Munich. On October 13, 1997 the New York Times reported on a number of examples of nuclear material smuggling from an insecure Russian system. The US has been assisting Russia in securing its fissile material under the Nunn-Lugar Program, but in recent years the government has been cutting funds for this.

Fissile Material Cut-off Treaty

A treaty banning fissile material has been on the agenda of the Conference on Disarmament (CD) for many years. However, differences in what it should cover have prevented negotiations. Some countries - including the NWS - wanted it to cover just the production of fissile material, while others - including Pakistan - wanted it to also address current stockpiles. Some states also want to see concurrent progress by the NWS on nuclear disarmament. There is also the question of how to deal with the production of non-fissile nuclear materials, especially tritium.

In 1998, some progress appeared possible when the CD established an ad hoc committee to discuss a proposed fissile material cut-off treaty. However, US plans to develop ballistic missile defence have added another damper on the situation. China hinted that it may increase its nuclear arsenal in response thus requiring more fissile material. Due to the difficulties in the CD, it may be preferable for existing moratoria on fissile material production by the NWS to be codified in a treaty negotiated outside the CD, thus not requiring support from all CD members.

Uranium

Uranium is a heavy metal which has been used as an abundant source of concentrated energy for over 60 years. Uranium occurs in most rocks in concentrations of 2 to 4 parts per million and is as common in the Earth's crust as tin, tungsten and molybdenum. Uranium occurs in seawater, and can be recovered from the oceans.

Uranium was discovered in 1789 by Martin Klaproth, a German chemist, in the mineral called pitchblende. It was named after the planet Uranus, which had been discovered eight years earlier. Uranium was apparently formed in supernovae about 6.6 billion years ago. While it is not common in the solar system, today its slow radioactive decay provides the main source of heat inside the Earth, causing convection and continental drift. The high density of uranium means that it also finds uses in the keels of yachts and as counterweights for aircraft control surfaces, as well as for radiation shielding. Uranium has a melting point of 1132 °C. The chemical symbol for uranium is U.

On a scale arranged according to the increasing mass of their nuclei, uranium is one of the heaviest of all the naturally-occurring elements (hydrogen is the lightest). Uranium is 18.7 times as dense as water.

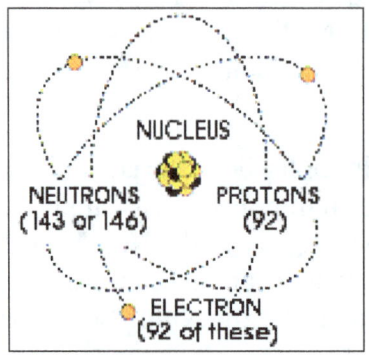

The Atom diagram.

Like other elements, uranium occurs in several slightly differing forms known as 'isotopes'. These isotopes differ from each other in the number of uncharged particles (neutrons) in the nucleus. Natural uranium as found in the Earth's crust is a mixture largely of two isotopes: uranium-238 (U-238), accounting for 99.3% and uranium-235 (U-235) about 0.7%.

The isotope U-235 is important because under certain conditions it can readily be split, yielding a lot of energy. It is therefore said to be 'fissile' and we use the expression 'nuclear fission'.

Meanwhile, like all radioactive isotopes, they decay. U-238 decays very slowly, its half-life being about the same as the age of the Earth (4500 million years). This means that it is barely radioactive, less so than many other isotopes in rocks and sand. Nevertheless it generates 0.1 watts/tonne as decay heat and this is enough to warm the Earth's core. U-235 decays slightly faster.

Energy from the Uranium Atom

The nucleus of the U-235 atom comprises 92 protons and 143 neutrons (92 + 143 = 235). When the nucleus of a U-235 atom captures a moving neutron it splits in two (fissions) and releases some energy in the form of heat, also two or three additional neutrons are thrown off. If enough of these expelled neutrons cause the nuclei of other U-235 atoms to split, releasing further neutrons, a fission 'chain reaction' can be achieved. When this happens over and over again, many millions of times, a very large amount of heat is produced from a relatively small amount of uranium.

It is this process, in effect 'burning' uranium, which occurs in a nuclear reactor. The heat is used to make steam to produce electricity.

Inside the Reactor

Nuclear power stations and fossil-fuelled power stations of similar capacity have many features in common. Both require heat to produce steam to drive turbines and generators. In a nuclear power station, however, the fissioning of uranium atoms replaces the burning of coal or gas. In a nuclear reactor the uranium fuel is assembled in such a way that a controlled fission chain reaction can be achieved. The heat created by splitting the U-235 atoms is then used to make steam which spins a turbine to drive a generator, producing electricity.

The chain reaction that takes place in the core of a nuclear reactor is controlled by rods which absorb neutrons and which can be inserted or withdrawn to set the reactor at the required power level.

The fuel elements are surrounded by a substance called a moderator to slow the speed of the emitted neutrons and thus enable the chain reaction to continue. Water, graphite and heavy water are used as moderators in different types of reactor.

Because of the kind of fuel used (i.e. the concentration of U-235), if there is a major uncorrected malfunction in a reactor the fuel may overheat and melt, but it cannot explode like a bomb.

A typical 1000 megawatt (MWe) reactor can provide enough electricity for a modern city of up to one million people.

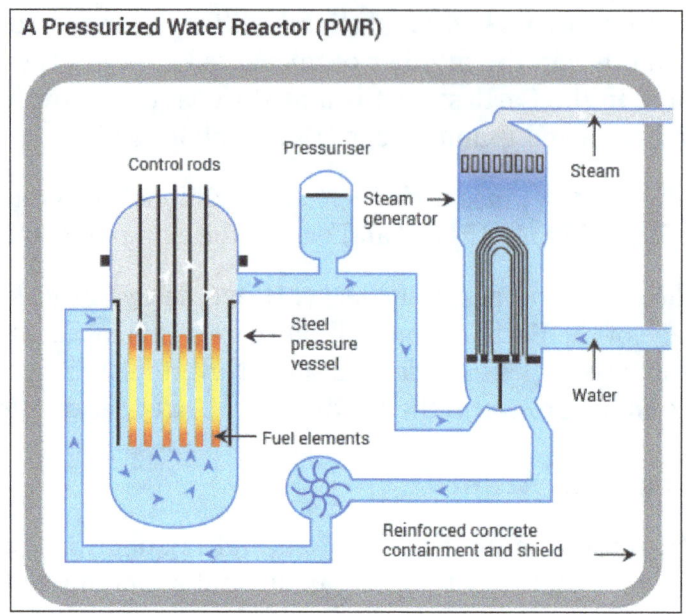

Uranium and Plutonium

Whereas the U-235 nucleus is 'fissile', that of U-238 is said to be 'fertile'. This means that it can capture one of the neutrons which are flying about in the core of the reactor and become (indirectly) plutonium-239, which is fissile. Pu-239 is very much like U-235, in that it fissions when hit by a neutron and this yields a similar amount of energy.

Because there is so much U-238 in a reactor core (most of the fuel), these reactions occur frequently, and in fact about one-third of the fuel's energy yield comes from 'burning' Pu-239.

But sometimes a Pu-239 atom simply captures a neutron without splitting, and it becomes Pu-240. Because the Pu-239 is either progressively 'burned' or becomes Pu-240, the longer the fuel stays in the reactor the more Pu-240 is in it. (The significance of this is that when the spent fuel is removed after about three years, the plutonium in it is not suitable for making weapons but can be recycled as fuel).

From Uranium Ore to Reactor Fuel

Uranium ore can be mined by underground or open-cut methods, depending on its depth. After mining, the ore is crushed and ground up. Then it is treated with acid to dissolve the uranium, which is recovered from solution.

Uranium may also be mined by in situ leaching (ISL), where it is dissolved from a porous underground ore body in situ and pumped to the surface.

The end product of the mining and milling stages, or of ISL, is uranium oxide concentrate (U_3O_8). This is the form in which uranium is sold.

Before it can be used in a reactor for electricity generation, however, it must undergo a series of processes to produce a useable fuel.

For most of the world's reactors, the next step in making the fuel is to convert the uranium oxide into a gas, uranium hexafluoride (UF_6), which enables it to be enriched. Enrichment increases the proportion of the uranium-235 isotope from its natural level of 0.7% to 4-5%. This enables greater technical efficiency in reactor design and operation, particularly in larger reactors, and allows the use of ordinary water as a moderator.

After enrichment, the UF_6 gas is converted to uranium dioxide (UO_2) which is formed into fuel pellets. These fuel pellets are placed inside thin metal tubes, known as fuel rods, which are assembled in bundles to become the fuel elements or assemblies for the core of the reactor. In a typical large power reactor there might be 51,000 fuel rods with over 18 million pellets.

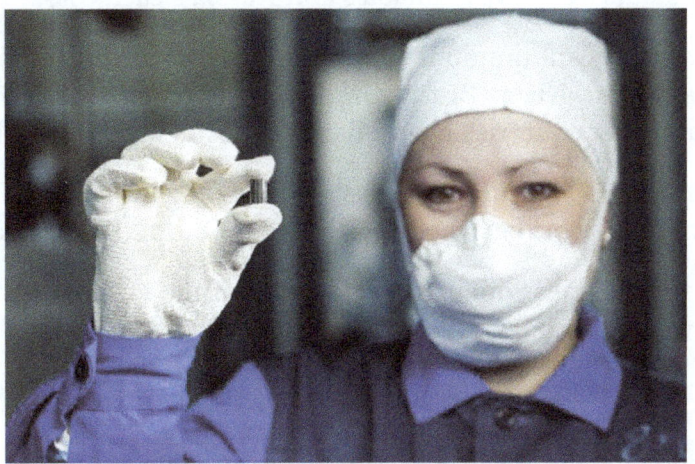

A worker holds up a newly made fuel pellet.

For reactors which use natural uranium as their fuel (and hence which require graphite or heavy water as a moderator) the U_3O_8 concentrate simply needs to be refined and converted directly to uranium dioxide.

When the uranium fuel has been in the reactor for about three years, the used fuel is removed, stored, and then either reprocessed or disposed of underground.

Who uses Nuclear Power?

About 11% of the world's electricity is generated from uranium in nuclear reactors. This amounts to over 2500 TWh each year, as much as from all sources of electricity worldwide in 1960.

It comes from over 440 nuclear reactors with a total output capacity of about 390,000 megawatts (MWe) operating in 31 countries. About 50 more reactors are under construction and over 100 are planned.

Belgium, Bulgaria, Czech Republic, Finland, France, Hungary, Slovakia, Slovenia, Sweden, Switzerland and Ukraine all get 30% or more of their electricity from nuclear reactors. The USA has just under 100 reactors operating, supplying 20% of its electricity. France gets over 70% of its electricity from uranium.

Over the 60 years that the world has enjoyed the benefits of cleanly-generated electricity from nuclear power, there have been over 17,000 reactor-years of operational experience.

Who has and who Mines Uranium?

Uranium is widespread in many rocks, and even in seawater. However, like other metals, it is seldom sufficiently concentrated to be economically recoverable. Where it is, we speak of an orebody. In defining what is ore, assumptions are made about the cost of mining and the market price of the metal. Uranium reserves are therefore calculated as tonnes recoverable up to a certain cost.

Australia's known resources are over 1.8 million tonnes of uranium recoverable at up to US$130/kg U (currently above the market 'spot' price), Kazakhstan's are over 800,000 tonnes of uranium and Canada's and Russia's are about 500,000 tU. Australia's resources in this category are 30% of the world's total, Kazakhstan's are 14%, Canada's and Russia's each 8%.

Several countries have significant uranium resources. Apart from the top four, they are in order: Namibia, South Africa, China, Niger, Brazil, Uzbekistan, Ukraine and Mongolia, all with 2% or more of the world total. Other countries have smaller deposits which could be mined if needed.

Kazakhstan is the world's top uranium producer, followed by Canada and then Australia as the main suppliers of uranium to world markets – now about 60,000 tU per year.

Uranium is sold only to countries which are signatories of the Nuclear Non-Proliferation Treaty (NPT), and which allow international inspection to verify that it is used only for peaceful purposes.

Other uses of Nuclear Energy

Many people, when talking about nuclear energy, have only nuclear reactors (or perhaps nuclear weapons) in mind. Few people realise the extent to which the use of radioisotopes has changed our lives over the last few decades.

Using relatively small special-purpose nuclear reactors, it is possible to make a wide range of radioactive materials (radioisotopes) at low cost. For this reason the use of artificially-produced radioisotopes has become widespread since the early 1950s, and there are now about 220 'research' reactors in 56 countries producing them. These are essentially neutron factories rather than sources of heat.

Radioisotopes

In our daily life we need food, water and good health. Today, radioactive isotopes play an important part in the technologies that provide us with all three. They are produced by bombarding small amounts of particular elements with neutrons.

In medicine, radioisotopes are widely used for diagnosis and research. Radioactive chemical tracers emit gamma radiation which provides diagnostic information about a person's anatomy and the functioning of specific organs. Radiotherapy also employs radioisotopes in the treatment of some illnesses, such as cancer. About one person in two in the Western world is likely to experience the benefits of nuclear medicine in their lifetime. More powerful gamma sources are used to sterilise syringes, bandages and other medical utensils – gamma sterilisation of equipment is almost universal.

In the preservation of food, radioisotopes are used to inhibit the sprouting of root crops after harvesting, to kill parasites and pests, and to control the ripening of stored fruit and vegetables. Irradiated foodstuffs are accepted by world and national health authorities for human consumption in an increasing number of countries. They include potatoes, onions, dried and fresh fruits, grain and grain products, poultry and some fish. Some prepacked foods can also be irradiated.

In the growing of crops and breeding livestock, radioisotopes also play an important role. They are used to produce high yielding, disease-resistant and weather-resistant varieties of crops, to study how fertilisers and insecticides work, and to improve the productivity and health of domestic animals.

Industrially, and in mining, they are used to examine welds, to detect leaks, to study the rate of wear of metals, and for on-stream analysis of a wide range of minerals and fuels.

There are many other uses. A radioisotope derived from the plutonium formed in nuclear reactors is used in most household smoke detectors.

Radioisotopes are used to detect and analyse pollutants in the environment, and to study the movement of surface water in streams and also of groundwater.

Other Reactors

There are also other uses for nuclear reactors. About 200 small nuclear reactors power some 150 ships, mostly submarines, but ranging from icebreakers to aircraft carriers. These can stay at sea for long periods without having to make refuelling stops. In the Russian Arctic where operating conditions are beyond the capability of conventional icebreakers, very powerful nuclear-powered vessels operate year-round, where previously only two months allowed northern access each year.

The heat produced by nuclear reactors can also be used directly rather than for generating electricity. In Sweden and Russia, for example, surplus heat is used to heat buildings. Nuclear heat may also be used for a variety of industrial processes such as water desalination. Nuclear desalination is likely to be a major growth area in the next decade.

High-temperature heat from nuclear reactors is likely to be employed in some industrial processes in future, especially for making hydrogen.

Military Sources of Fuel

Both uranium and plutonium were used to make bombs before they became important for making electricity and radioisotopes. The type of uranium and plutonium for bombs is different from that in a nuclear power plant. Bomb-grade uranium is highly-enriched (>90% U-235, instead of up to 5%); bomb-grade plutonium is fairly pure Pu-239 (>90%, instead of about 60% in reactor-grade) and is made in special reactors.

Since the 1990s, due to disarmament, a lot of military uranium has become available for electricity production. The military uranium is diluted about 25:1 with depleted uranium (mostly U-238) from the enrichment process before being used in power generation. Over two decades to 2013 one-tenth of US electricity was made from Russian weapons uranium.

Plutonium

Plutonium-239 is one of the two fissile materials used for the production of nuclear weapons and in some nuclear reactors as a source of energy. The other fissile material is uranium-235. Plutonium-239 is virtually nonexistent in nature. It is made by bombarding uranium-238 with neutrons in a nuclear reactor. Uranium-238 is present in quantity in most reactor fuel; hence plutonium-239 is continuously made in these reactors. Since plutonium-239 can itself be split by neutrons to release energy, plutonium-239 provides a portion of the energy generation in a nuclear reactor.

Table: Physical Characteristics of Plutonium Metal.

Color	Silver
Melting point	641 °C
Boiling point	3232 °C
Density	16 to 20 grams/cubic centimeter

Plutonium belongs to the class of elements called transuranic elements whose atomic number is higher than 92, the atomic number of uranium. Essentially all transuranic materials in existence are manmade. The atomic number of plutonium is 94.

Plutonium has 15 isotopes with mass numbers ranging from 232 to 246. Isotopes of the same element have the same number of protons in their nuclei but differ by the number of neutrons. Since the chemical characteristics of an element are governed by the number of protons in the nucleus, which equals the number of electrons when the atom is electrically neutral (the usual elemental form at room temperature), all isotopes have nearly the same chemical characteristics. This means that in most cases it is very difficult to separate isotopes from each other by chemical techniques.

Only two plutonium isotopes have commercial and military applications. Plutonium-238, which is made in nuclear reactors from neptunium-237, is used to make compact thermoelectric generators; plutonium-239 is used for nuclear weapons and for energy; plutonium-241, although fissile, is impractical both as a nuclear fuel and a material for nuclear warheads. Some of the reasons are far higher cost , shorter half-life, and higher radioactivity than plutonium-239. Isotopes of plutonium with mass numbers 240 through 242 are made along with plutonium-239 in nuclear reactors, but they are contaminants with no commercial applications. In this fact sheet we focus on civilian and military plutonium, which consist mainly of plutonium-239 mixed with varying amounts of other isotopes, notably plutonium-240, -241, and -242.

Plutonium-239 and plutonium-241 are fissile materials. This means that they can be split by both slow (ideally zero-energy) and fast neutrons into two new nuclei (with the concomitant release of energy) and more neutrons. Each fission of plutonium-239 resulting from a slow neutron absorption results in the production of a little more than two neutrons on the average. If at least one of these neutrons, on average, splits another plutonium nucleus, a sustained chain reaction is achieved.

The even isotopes, plutonium-238, -240, and -242 are not fissile but yet are fissionable—that is, they can only be split by high energy neutrons. Generally, fissionable but non-fissile isotopes cannot sustain chain reactions; plutonium-240 is an exception to that rule.

The minimum amount of material necessary to sustain a chain reaction is called the critical mass. A supercritical mass is bigger than a critical mass, and is capable of achieving a growing chain reaction where the amount of energy released increases with time.

The amount of material necessary to achieve a critical mass depends on the geometry and the density of the material, among other factors. The critical mass of a bare sphere of plutonium-239 metal is about 10 kilograms. It can be considerably lowered in various ways.

The amount of plutonium used in fission weapons is in the 3 to 5 kilograms range. According to a recent Natural Resources Defense Council report, nuclear weapons with a destructive power of 1 kiloton can be built with as little as 1 kilogram of weapon grade plutonium. The smallest theoretical critical mass of plutonium-239 is only a few hundred grams.

In contrast to nuclear weapons, nuclear reactors are designed to release energy in a sustained fashion over a long period of time. This means that the chain reaction must be controlled—that is, the number of neutrons produced needs to equal the number of neutrons absorbed. This balance is achieved by ensuring that each fission produces exactly one other fission.

All isotopes of plutonium are radioactive, but they have widely varying half-lives. The half-life is the time it takes for half the atoms of an element to decay. For instance, plutonium-239 has a half-life of 24, 110 years while plutonium-241 has a half-life of 14.4 years. The various isotopes also have different principal decay modes. The isotopes present in commercial or military plutonium-239 are plutonium-240, -241, and -242. Table shows a summary of the radiological properties of five plutonium isotopes.

The isotopes of plutonium that are relevant to the nuclear and commercial industries decay by the emission of alpha particles, beta particles, or spontaneous fission. Gamma radiation, which is penetrating electromagnetic radiation, is often associated with alpha and beta decays.

Table: Radiological Properties of Important Plutonium Isotopes.

	Pu-238	Pu-239	Pu-240	Pu-241	Pu-42
Half-life (in years)	87.74	24,110	6537	14.4	376,000
Specific activity (curies/gram)	17.3	0.063	0.23	104	0.004
Principal decay mode	alpha	alpha	alpha (some spontaneous fission	beta	alpha
Decay energy (MeV)	5.593	5.244	5.255	0.021	4.983
Radiological hazards	alpha, weak gamma	alpha, weak gamma	alpha, weak gamma	beta, weak gamma	alpha, weak gamma

Chemical Properties and Hazards of Plutonium

Table describes the chemical properties of plutonium in air. These properties are important because they affect the safety of storage and of operation during processing of plutonium. The oxidation of plutonium represents a health hazard since the resulting stable compound, plutonium dioxide is in particulate form that can be easily inhaled. It tends to stay in the lungs for long periods, and is also transported to other parts of the body. Ingestion of plutonium is considerably less dangerous since very little is absorbed while the rest passes through the digestive system.

Table: How Plutonium Metal Reacts in Air.

Forms and Ambient Conditions	Reaction
Non-divided metal at room temperature (corrodes)	relatively inert, slowly oxidizes
Divided metal at room temperature (PuO_2)	readily reacts to form plutonium dioxide
Finely divided particles under about 1 millimeter diameter	>spontaneously ignites at about 150 °C
Finely particles over about 1 millimeter diameter	spontaneously ignites at about 500 °C.
Humid, elevated temperatures (PuO_2)	readily reacts to form plutonium dioxide

Important Plutonium Compounds and their Uses

Plutonium combines with oxygen, carbon, and fluorine to form compounds which are used in the nuclear industry, either directly or as intermediates.

Table shows some important plutonium compounds. Plutonium metal is insoluble in nitric acid and plutonium is slightly soluble in hot, concentrated nitric acid. However, when plutonium dioxide and uranium dioxide form a solid mixture, as in spent fuel from nuclear reactors, then the solubility of plutonium dioxide in nitric acid is enhanced due to the fact that uranium dioxide is soluble in nitric acid. This property is used when reprocessing irradiated nuclear fuels.

Table: Important Plutonium Compounds and Their Uses.

Compound	Use
Oxides Plutonium Dioxide (PuO_2)	Can be mixed with uranium dioxide (UO_2) for use as reactor fuel
Carbides Plutonium Carbide (PuC) Plutonium Dicarbide (PuC_2) Diplutonium Tricarbide (Pu_2C_3)	All three carbides can potentially be used as fuel in breeder reactors
Fluorides Plutonium Trifluoride (PuF_3) Plutonium Tetrafluoride (PuF_4)	Both fluorides are intermediate compounds in the production of plutonium metal
Nitrates Plutonium Nitrates $[Pu(NO_3)_4]$ and $[Pu(NO_3)_3]$	No use, but it is a product of reprocessing (extraction of plutonium from used nuclear fuel)

Formation and Grades of Plutonium-239

Plutonium-239 is formed in both civilian and military reactors from uranium-238.

The subsequent absorption of a neutron by plutonium-239 results in the formation of plutonium-240. Absorption of another neutron by plutonium-240 yields plutonium-241. The higher isotopes are formed in the same way. Since plutonium-239 is the first in a string of plutonium isotopes created from uranium-238 in a reactor, the longer a sample of uranium-238 is irradiated, the greater the percentage of heavier isotopes. Plutonium must be chemically separated from the fission products and remaining uranium in the irradiated reactor fuel. This chemical separation is called reprocessing.

Fuel in power reactors is irradiated for longer periods at higher power levels, called high "burn-up", because it is fuel irradiation that generates the heat required for power production. If the goal is production of plutonium for military purposes then the "burn-up" is kept low so that the plutonium-239 produced is as pure as possible, that is, the formation of the higher isotopes, particularly plutonium-240, is kept to a minimum.

Plutonium has been classified into grades by the US DOE (Department of Energy) as shown in table.

Table: Grades of Plutonium.

Grades	Pu-240 Content
Supergrade	2-3%
Weapon grade	
Fuel grade	7-19%
Reactor grade	19% or greater

It is important to remember that this classification of plutonium according to grades is somewhat arbitrary. For example, although "fuel grade" and "reactor grade" are less suitable as weapons material than "weapon grade" plutonium, they can also be made into a nuclear weapon, although the yields are less predictable because of unwanted neutrons from spontaneous fission. The ability of countries to build nuclear arsenals from reactor grade plutonium is not just a theoretical construct. It is a proven fact. During a June 27, 1994 press conference, Secretary of Energy Hazel O'Leary revealed that in 1962 the United States conducted a successful test with "reactor grade" plutonium. All grades of plutonium can be used as weapons of radiological warfare which involve weapons that disperse radioactivity without a nuclear explosion.

Thorium

Since the beginning of the nuclear era, thorium has been acknowledged as an interesting resource for its potential use as nuclear fuel. In the early period of nuclear energy, thorium had been considered as a possible supplement or even a replacement for uranium which was feared to be scarce at the time. Thorium is in all likelihood relatively abundant on earth and presents a number of intrinsic nuclear and chemical properties that would make its use as a potential nuclear fuel particularly interesting. However, in the early years of nuclear energy, it was soon discovered that the supply of natural uranium was not as limited as initially projected. Moreover, thorium lacks a fissionable isotope – a major drawback for thorium, as it is impossible to start a fission chain reaction purely on natural thorium – and consequently any nuclear system using thorium would initially be dependent on prior generation of fissile matter (extracted *from* uranium or bred *in* uranium systems). A uranium-plutonium fuel cycle was therefore not only an easier but a necessary first step, in line with the strategies of the main countries developing nuclear energy at the time.

Extensive reviews of thorium use in nuclear reactors have been published in the literature over the years. From 1950 to the late 1970s, the thorium fuel cycle was the subject of numerous studies and pilot experiments. Power reactors were operated with thorium-based fuels, demonstrating the feasibility as well as the complexities associated with their use.

In the late 1970s and 1980s, public support for nuclear power declined after the Three Mile Island accident in April 1979 and the Chernobyl accident seven years later in 1986. In conjunction with these events, the price of uranium fell to very low levels in the 1980s and thus the search for an alternative to uranium became economically of little interest. Also, the cessation of efforts to deploy commercial spent fuel reprocessing in the United States in the late 1970s – motivated primarily by the perceived potential proliferation risks associated with the separation of fissile materials – ended interest in closed nuclear fuel cycles in the United States. This also had an impact on decision making in other countries.

In the case of thorium cycles, the decision to end reprocessing removed the possibility of recovering the ^{233}U from irradiated fuel in the United States. In other countries, such as France, the uranium-plutonium fuel cycle was under full development, with the implementation of partial recycling and no impetus to change from the uranium-plutonium fuel cycle. The thorium option was, however, never abandoned and has continued to be studied with fluctuating intensity, particularly in academia. Today, the availability of fissile material (plutonium or enriched uranium) arising from the well-established uranium plutonium fuel cycle makes the implementation of thorium fuels feasible in principle, although the necessary economic drivers for devoting significant industrial resources to that end have not yet been clearly established.

The 2015 Nuclear Energy Agency (NEA) report Introduction of Thorium in the Nuclear Fuel Cycle identified general conditions under which a transition to a thorium fuel cycle would become a practical option, providing details of the technical challenges associated with the various stages and options during that transition.

Renewed Interest in Thorium

Today, many countries are developing or exploring the construction of new nuclear power plants. While several factors, including the financial crisis, the Fukushima Daiichi accident in March 2011 and the natural gas boom in North America, have impacted plans to build new plants in several countries, future energy scenarios continue to project "a significant development of nuclear energy to meet energy and environmental goals, albeit at a somewhat slower rate than previously projected". The arguments behind these projections are based on the importance of issues including:

- LOw carbon energy production;
- Security of energy supply;
- Availability of competitive baseload electrical supply;
- Other potential uses of nuclear energy, such as process heat generation.

On the other hand, issues related mainly to spent fuel management from the current uranium-plutonium cycle, still present challenges. These challenges include the:

- Deployment of geological repositories for direct disposal of intact spent fuel and disposal of vitrified waste;

- Development of uranium-plutonium fast neutron reactors (FRs);

- Transition to generation IV systems.

In some countries, concern over the eventual disposal of spent nuclear fuel has grown. Delays in repository programmes have meant that most countries will have to face an accumulation of spent fuel and high-level waste. Drivers for thorium fuel development Given thorium's potential for reduced minor actinide production, its introduction into the fuel cycle continues to be given consideration. In the last fifteen years, increased focus has been given to research on thorium-based fuels for medium-term utilisation in present or evolutionary (generation III+) reactors so as to provide additional options in the management of uranium and plutonium.

Thorium could potentially play a useful role as a complement to the uranium plutonium fuel cycle as it would enhance the management of used fuels and radioactive waste while also providing a means of dealing with stockpiles of plutonium in the absence of fast reactors. Additionally, thorium could provide flexibility within the context of uncertainty around the long-term availability of relatively cheap uranium. Studies are being carried out in several areas, aiming at quantifying the potential benefits related to thorium fuels. More specific areas of interest where thorium could play a favourable role in the future are:

- Improvements in the utilisation and management of fissile materials: Improvements in the current and in future fuel cycles will open up possibilities from the short term onwards. Of particular interest is the possibility of reaching higher conversion factors or even breeding conditions in thermal or epithermal neutron spectra evolutionary generation III+ systems that use thorium-based fuels, with the aim of recovering the fissile material from the used fuel.

- Fuel performance: Thorium has very promising physicochemical characteristics that are good candidates for improving fuel performance, in particular in the form of thorium dioxide, which has a high thermal conductivity, a low thermal expansion coefficient and a high melting temperature.

- Waste management: Thorium fuels may lead to less minor actinide production per unit of energy produced, although this depends on the fissile seed used. Thorium may be used as an inert matrix in view of burning plutonium (and possibly other actinide elements) which could provide an option for plutonium management. In the long term, $Th/^{233}U$ fuel cycles would lead to less minor actinides in the waste streams. It should be recognized, however, that there remain considerable uncertainties associated with the practical application of thorium. With the lack of clear economic incentives to deploy thorium, the industrial development of thorium as a replacement for uranium in the fuel cycle is likely to remain limited.

Fertile-fissile Cycles

Today, water-moderated reactor technologies are dominant around the world. According to IAEA statistics, of the 437 power reactors in operation in 2013, more than 90% were water moderated: 273 were pressurised water reactors (PWRs), 84 boiling water reactors (BWRs) and 48 pressurised heavy water reactors (PHWRs). These technologies rely on the extraction of energy released by fissions triggered by thermalised neutrons.

Heavy nuclides that fission under thermal neutron interaction are called "fissile". Others may be called "fertile" because they are more likely to absorb a thermal neutron instead of undergoing fission and thereby transmute into a heavier nuclide, which may be fissile.

When irradiated by thermal neutrons, ^{232}Th and ^{238}U follow similar processes. ^{232}Th breeds ^{233}U in a completely analogous way to that in which ^{238}U breeds ^{239}Pu. These two processes of neutron "radiative capture" (n, γ) reactions, followed by consecutive radioactive decays (β-) represent the two practical "fertile/fissile fuel cycles"; the ^{238}U/^{239}Pu fuel cycle and the ^{232}Th/^{233}U fuel cycle.

$$^{238}\text{Th} \xrightarrow{(n,\gamma)} {}^{233}\text{Th} \xrightarrow[T_{1/2}=22\min]{\beta^-} {}^{239}\text{Np} \xrightarrow[T_{1/2}=27d]{\beta^-} {}^{233}\text{U}$$

$$^{238}\text{U} \xrightarrow{(n,\gamma)} {}^{239}\text{U} \xrightarrow[T_{1/2}=23.5\min]{\beta^-} {}^{239}\text{Np} \xrightarrow[T_{1/2}=2.3d]{\beta^-} {}^{239}Pu$$

Conversion Ratios and Breeding

Although it requires an initial fissile "seed", a thorium fuel cycle may ultimately become autonomous in terms of fissile inventory as creation of the fissile nuclide 233U occurs in fertile thorium, provided the retrieval of this 233U is viable through reprocessing and separation. Under certain conditions, it is possible to create (or breed) more fissile material than is consumed by the reactor.

The conversion ratio (CR), also called the breeding ratio, is defined as the ratio of the rate of production to the rate of consumption of fissile nuclei in the fuel of a reactor in operation. If CR is greater than or equal to 1, the system is called a "breeder". Reaching a conversion ratio of allows a reactor in a closed fuel cycle to function independently from any continuous external supply of fissile matter (assuming negligible losses at the reprocessing stage).

Thorium Resources

Thorium is a common natural element and fairly evenly spread in the crust of the earth. The main mineral host for thorium is monazite, a rare earth phosphate generally containing between 6 to 12% thorium-oxide. Countries possessing significant amounts of thorium resources include Australia, Brazil, Egypt, India, Norway, Russia, Turkey and the United States.

The latest estimates in the joint report by the NEA and the IAEA on Uranium: Resources, Production and Demand identified thorium resources amounting to about 6.2 million tonnes. Some studies, however, such as the one published by the United States Geological Survey, provide lower estimates of thorium resources of around 1-2 million tonnes. There is currently no standard classification for thorium resources and these identified resources do not have the same meaning in terms of classification as identified uranium resources.

Availability of Thorium

To date, thorium production has mainly emerged as a by-product of mining activities for rare earth elements and uranium due to its association with these elements in nature. Under the present conditions, dedicated extraction and production of thorium is not considered economically viable due to the current lack of a thorium demand and market. However, such a dedicated extraction would probably not be needed in the short to medium term, because the increase of demand in the past decade for rare earth ores (mainly used in electronic equipment) has resulted in a by-product supply path of thorium ores that is currently independent of the demand for thorium from the nuclear market. Currently, demand for thorium is mainly accounted for by non-energy related companies, which use this material for applications such as high-temperature ceramics, melting tanks, catalysts, welding electrodes and some alloys.

Furthermore, many of the world's monazite deposits, such as the thorium-recovering sites of India, are co-located with titanium ore (ilmenite). Thorium by-product recovery from rare earth elements, as well as from currently active titanium mines, could potentially be greater than the estimated volumes of uranium consumed yearly by the world's entire nuclear reactor fleet.

Mining

Thorium is generally easier to mine than uranium. One of the advantages of thorium mining is that thorium can be extracted from open pit monazite deposits (presently, the main source of thorium), which are easier to mine than most uranium-bearing ores. Management of thorium mine tailings is also simpler because of the much shorter halflife of one of its daughter products, 220Rn (55 sec), compared to the equivalent daughter product of uranium, ^{222}Rn (3.8 days). However, the radioactivity of the mined products is much higher for thorium than for uranium, because of the thorium decay chain product ^{208}Tl, which emits 2.6 MeV gamma rays.

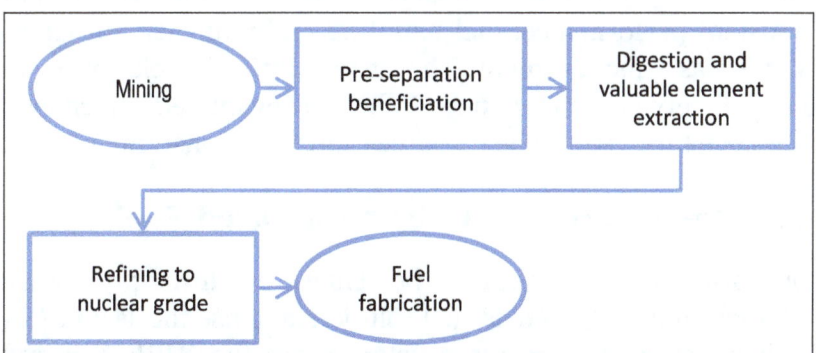

Overview of theorium recovery process.

The uranium mining industry produces large amounts of unused depleted uranium, which is mostly 238U. If a thorium mining industry were to be developed, the volumes of thorium needed to be mined would in all likelihood be significantly lower than the volumes involved for mineral uranium today, and even lower in the case of higher conversion or breeding cycles that optimise the use of thorium.

Fuel Cycle Aspects

Fuel Fabrication and Testing

The technical feasibility of using thorium as a fuel component in current power reactors has been shown in numerous studies. Thorium dioxide (thoria) has material properties (low-thermal expansion coefficient, high-thermal conductivity and high-melting temperature) that make it well-suited for use as a fertile fuel matrix in present reactors and for consuming plutonium or transmuting transuranic nuclides, especially as compared with the uranium dioxide currently used in MOX fuels. Thoria-based fuels for light water reactors (LWRs) and heavy water reactors (HWRs) show a potential for improved fuel performance in terms of reduced fission product release and reduced erosion, which lead to fewer fuel defects. Test irradiation programmes of thoria fuelshave been carried out in the past, are currently on-going or are foreseen to determine the key properties and behaviour of thorium dioxide fuels, such as thermal conductivity, swelling and fission gas release as a function of burn-up. The results constitute a significant step towards the broader use of thoria fuel ceramics in existing reactors.

Thorium-based fuels would need to be properly qualified to assure their safe performance in the usual suite of normal and accident scenarios of prime concern to regulators. Although no major obstacles have been found thus far, processes have to be further developed to manufacture thorium-based fuels at an industrial scale.

Performance of newly developed fuels would then need to be confirmed in further test programmes in agreement with licensing authorities, taking into consideration today's heightened safety requirements. The introduction of thorium in a reactor core would most likely follow a progressive approach – single fuel rod irradiation tests, followed by irradiation of test assemblies, leading to the introduction of thorium fuel in several assemblies or reload batches in the core.

A recent report by the United States Nuclear Regulatory Commission (NRC) on the Safety and Regulatory Issues of the Thorium Fuel Cycle, highlights that thorium's "fundamental nuclear properties have impacts on a number of key areas related to reactor and safety analyses, including steady state and transient performance, fuel handling and management (fresh and irradiated), reactor operations and waste management. The uncertainties in relation to these data and the resulting impact on key safety parameters need to be fully evaluated". Such evaluation processes, even for existing infrastructures, would be resource- and time-consuming.

Reprocessing and Re-fabrication of Thorium-based Fuels

The reprocessing of thorium-based fuel cannot be achieved with the processes that are currently used for uranium-plutonium fuel. The envisaged solution is to use the THOREX (thorium recovery by extraction) liquid-liquid extraction process, developed from the PUREX (Pu and U recovery by extraction) process concepts used for uranium-plutonium fuel. The THOREX process (and its variants) has been successfully used at pilot scale, notably in the United States to reprocess about 900 tonnes of thorium fuel (~1.5 tonnes ^{233}U) and also in France. However, the dissolution of thorium metal or oxide has proven to be more complex than for uranium and requires the addition of hydrofluoric and nitric acids. The corrosiveness of these acids with respect to the process equipment must be buffered by adding aluminium nitrate. Such specific processes are yet to be developed an industrial scale and would ultimately require further development of the dissolution, the liquid-liquid extraction and the conversion steps. The influence of the burn-up of the fuel and the impact of the conversion and fabrication processes, as well as the fuel microstructure, would also require further investigation.

Different Options for Thorium Implementation

While the longer-term options of thorium use aim explicitly at the replacement of uranium-plutonium in a "pure" Th/^{233}U fuel cycle, such solutions would need to transit through a period where thorium, together with the ^{233}U produced, is gradually introduced into the fuel cycle.

A full transition to a breeder thorium fuel cycle is undoubtedly a long-term process. India, a country that for strategic reasons of energy independence has made great efforts at a national level to fully develop and deploy the thorium fuel cycle, does not envisage reaching a self-sustained Th/^{233}U cycle before 2070. The three-stage development programme supported by India is an example of the long-term staging of thorium introduction, limited not only by fissile material availability but also by the level of readiness of thorium breeder technologies.

It is important to differentiate the different categories of potential thorium implementation in reactors, both in the short (before 2030), medium (2030-2050) and longer term (post-2050). These different uses of thorium share very different development challenges and advantages.

Short Term

Thorium as an additive to the uranium-plutonium fuel cycle. In the short term (short by industrial timescales, i.e. before 2030), the introduction of thorium in small quantities as an "additive" in uranium-plutonium fuels can be considered as a practical option. A typical 5% to 10% mass fraction content of thoria in uraniumplutonium oxide fuels could be a means to improve the neutronic characteristics of PWR fuel assemblies, allowing for better core power flattening as well as a reduction in the use of burnable poisons such as gadolinium.

While this use of thorium surely does not address global objectives such as uranium savings, fissile material management or improvements in ultimate waste management, it may well be part of an initial step towards thorium use in generation-III or III+ reactors and could augment the irradiation experience towards higher burn-ups with thorium in LWRs or HWRs.

Medium Term

Thorium in generation III or III+ reactors as a complement to the uranium fuel cycle. Most likely over the next 20 years, new nuclear power plants under construction will nearly all be water-moderated technologies designed to last for about 60 years. Light water reactors will therefore continue to dominate nuclear energy production for most of the 21st century, with heavy water reactors also present in the market and showing promise as potential hosts for thorium fuels.

A 2011 NEA report on Trends towards Sustainability in the Nuclear Fuel Cycle concludes that "the successful large-scale reactor technology demonstration efforts conducted in the past suggest that there should not be insurmountable technical obstacles preventing the use of thorium fuel and its fuel cycle in the existing and evolutionary LWRs. However, the industrial infrastructure, research, design and licensing data are not in place to allow a rapid deployment of thorium fuels in current reactors in the short term".

A medium-term (2030-2050) option would be to use thorium in existing systems – in either LWRs (i.e. PWRs, BWRs) or HWRs (i.e. CANDUs). Thorium use could be considered, for example, both in a homogeneous and a heterogeneous manner through the following means:

- Mixing thorium with uranium and plutonium oxides as a homogeneous fuel;

- Using fuel assemblies with a combination of uranium and plutonium oxide fuel rods and thorium oxide fuel rods;

- Using a heterogeneous core approach, with the introduction of thorium fuel assemblies separated from the rest of the assemblies (UOX or MOX).

In the case of existing reactors, thorium would likely be used either as pure thoria in heterogeneous assemblies or cores, or with different combinations of thorium in homogeneously mixed oxide form with either low enriched uranium (LEU, enriched uranium containing less than 20% of the isotope ^{235}U), or plutonium (of adequate isotopic quality). Highly enriched uranium (HEU, enriched uranium containing 20% or more of the isotope ^{235}U) or pure ^{235}U would generally not be considered because of the worldwide consensus on the ban of their use for proliferation reasons.

Various studies have indicated that a workable approach for thorium use in LWRs could be via the heterogeneous option. Given the not-yet-proven industrial feasibility of mixed thorium-uranium-plutonium reprocessing schemes, the heterogeneity option would initially be at the level of fuel assembly, where the thorium and ^{233}U are part of the uranium-plutonium cycle, but without the development of a dedicated ^{233}U reprocessing/re-fabrication route.

Because of the current lack of reprocessing facilities and industrial processes for reprocessing thorium fuels, any near- to medium-term use of thorium fuels will likely be implemented in a once-through mode. In general, once-through fuel cycles using thorium would aim at achieving increased burn-up of the fuel, since the ^{233}U will be burnt in situ rather than recovered. In this respect, evolutionary generation III+ variants of PWRs, BWRs and HWRs could take advantage of such an improved utilisation of fissile resources.

Towards Higher Conversion

Hardening the neutron spectrum in the reactor is the principle behind high conversion in reduced-moderation generation III+ evolutionary variants of PWRs, BWRs and HWRs.

These concepts have implemented modifications to their respective base designs that, while not allowing them to achieve breeding ratios of 1, could nevertheless lead to higher conversion ratios using thorium-based fuels, due to their inherently favourable neutron economy in thermal or epithermal neutron spectra. 233U bred by thorium – which has the highest number of spare neutrons available for breeding at these neutron energies – is well-suited to these evolutionary designs, notably in the slightly modified, tighter fuel lattices of HWRs or BWRs such as modified CANDUs or potentially in resource-renewable BWR designs (RBWRs).

Even if high conversion does not reach breeding, achieving higher conversion ratios would still be desirable in view of the better utilisation of natural resources. A rapidly growing generation III or III+ nuclear reactor fleet could thus benefit from the use of thorium in parallel to uranium so as to improve their overall fissile material balance and the management of spent fuel through (multi-) recycling of uranium, and particulary plutonium. These types of strategies could become attractive options in cases where fast neutron reactor deployment, in synergy with generation III reactors, does not occur, is not sufficiently deployed or is compromised for any reason.

This potential can be further increased if multi-recycling of thorium fuel is implemented. Schemes with thorium facilitating the recycling of reprocessed uranium in CANDUs, or improving the recyclability of plutonium in LWRs, are under study and are deemed realisable in the medium term provided reprocessing challenges can be met.

Thorium-plutonium fuels in mixed oxide forms (Th-MOX) have long been considered a good option for plutonium disposal, as no significant quantities of new "second-generation" plutonium are produced through neutron irradiation of thorium. It should be noted, however, that the presence of MOX could, in some scenarios, result in an increase of the production of transuranic elements in the fuel, particularly compared to UOX systems. Nevertheless, Th-MOX provides a credible option for recycling of plutonium in the medium to longer term, provided the reprocessing challenges are met at an industrial level.

Long Term: Dedicated Breeder Systems using Th/^{233}U Closed Fuel Cycles

Longer-term options (i.e. post-2050) for thorium implementation may investigate an increased use of thorium by using combinations of reactor types or dedicated breeder reactor systems to establish a full Th/^{233}U fuel cycle. All these longer-term options would require fissile material — either plutonium of adequate fissile quality from the existing uranium-plutonium fuel cycle or ^{233}U bred during a transitioning period — in order to be feasible.

All advanced applications of thorium as nuclear fuel would require significant research and development in thorium-based fuel technology, as well as proper qualification of the fuel. An autonomous thorium fuel cycle is only possible if the ^{233}U is recycled. Advanced fuel cycle options in particular will depend on successful development of processes associated with spent thorium fuel reprocessing and re-fabrication of irradiated fuels with ^{233}U, including appropriate consideration of radiation protection and nonproliferation issues. These options would also depend on the successful implementation of the THOREX reprocessing method at industrial scales.

Generation-IV Concepts with Thorium

The Generation IV International Forum (GIF) has chosen molten salt reactors (MSRs) as one of its six concepts; the only one that specifically considers the use of thorium. MSR concepts implement very innovative fuel management approaches with the use of fuel (thorium- or uranium-based) in liquid form, which in principle allows continuous "online" reprocessing of the fuel in order to extract fission products and ^{233}Pa (the precursor of ^{233}U and a neutron absorber). Liquid fuel concepts, in principle, allow greater power densities and smaller initial fissile inventories than solid fuel concepts. The online liquid fuel management of MSRs allows for theoretical breeding ratios equal or greater than one.

In its latest Technology Roadmap Update for Generation IV Nuclear Energy Systems, GIF has extended the viability study phase of MSR concepts until 2025, reflecting the fact that MSR concepts are still in need of substantial development before they are deemed technologically feasible. Particularly challenging is the essential step of the online treatment of liquid fuel, which requires the implementation of pyro-chemical processes. Much, however, remains unknown about the actual feasibility of these processes or performances. In addition, safety analysis methods in their current form cannot be applied to liquid-fuelled MSRs, because of the innovative form of the fuel (i.e. absence of cladding, molten fuel conditions under normal operation, continuous circulation of fuel in and out of the active core). Development of methodologies for the design and safety evaluations of liquid-fuelled MSRs is also necessary.

In 2011, China announced the start of an ambitious R&D programme on molten salt reactors led by the Chinese Academy of Sciences (CAS) aiming, in particular, at the construction of a 2-MW pilot thorium molten salt reactor-liquid fuel (TMSR-LF). In 2014, the TMSR-LF prototype was in a pre-conceptual design phase with limited use of thorium in the foreseen candidate compositions for the molten salt fuel.

Hybrid Reactor Concepts

Other, even more innovative hybrid reactor concepts that combine the characteristics of different future reactor concepts – namely, hybrids of accelerator-driven systems with molten salt reactor systems, hybrids of fission and fusion reactors – have been envisaged as potentially making use of thorium. Although these concepts may have interesting theoretical properties, they inevitably reflect the disadvantages, uncertainties and unknown factors related to the various base technologies that compose them. These unknown elements are often independent from the fact that the concepts may or may not use thorium and, as such, would first need to be further studied, developed and demonstrated. Consequently, such composite or "hybrid" concepts are very unlikely to provide any credible application for commercial electricity production in this century.

Waste Management Aspects

The use of thorium as nuclear fuel is often associated with advantages in the radio toxicity of the resulting waste as compared to conventional uranium fuels. It must not be overlooked, however, that the implementation of thorium fuels with the view of developing a self-sustainable thorium fuel cycle will require the use of mixed fuel forms (thorium-LEU or thorium-plutonium fuels) during very long transition phases before a full Th/^{233}U cycle can be achieved. For these mixed fuel forms, a comparison in terms of advantages or disadvantages over current UO_2 fuels will strongly depend on the mixed fuel form considered and on fuel management and recycling strategies.

While a pure Th/^{233}U cycle will indeed produce considerably less plutonium and minor actinides than conventional UO_2 fuels, this is not the case for thorium-plutonium mixed fuel forms, and is less clear for thorium-LEU fuels. Furthermore, decay products from ^{233}U drive radiotoxicity to a higher level than that of LEU or U/Pu for the period between about ten thousand years and one million years, primarily due to the presence of ^{234}U (mainly produced through neutron capture on ^{233}U) and its decay product, 226Ra.

The relative differences between radiotoxicities resulting from the use of both cycles vary greatly depending on the recycling strategies and recycling efficiencies considered and must therefore be interpreted with care.

Irradiation of thorium fuel also gives rise to specific long-lived heavy radionuclides, some of which have a more important radiotoxicity than their counterparts in the uranium cycle. The most prominent are:

- ^{233}U with a half-life of about 160 000 years. It has a daughter product (^{229}Th), which contributes significantly to the radiotoxicity of the spent fuel. This difference becomes much less noticeable if ^{233}U is recycled.

- ^{231}Pa with a half-life of 33 000 years. It has a particularly high radiotoxicity, even higher than the principal isotopes of plutonium (^{239}Pu and ^{240}Pu).

- ^{233}U with a half-life of 70 years. The main burden comes from its decay products (212Bi and 208Tl), which are highly radiant (gamma radiation of 1.6 and 2.6 MeV) and which would impose heavy shielding requirements in recycling plants.

In terms of the potential reduction of the radiotoxic inventory of the waste compared to uranium-plutonium fuels, the benefits of a self-sustaining Th/^{233}U fuel cycle at equilibrium are more modest when considering the transition needed to establish that equilibrium cycle. The long-term radiotoxicity of thorium-based spent nuclear fuels is more accurately described as being comparable to that of uranium-based spent nuclear fuels.

Non-proliferation Aspects

The non-proliferation of nuclear weapons has always been an essential consideration to take into account in the deployment of any nuclear energy technology or process. This issue was addressed in particular in the framework of an extensive study conducted between 1978 and 1980 on the overall nuclear fuel cycle. A general conclusion reached in this report was that the technical obstacles to military use of thorium cycles with uranium enriched to less than 20% are similar to those of the uranium-plutonium cycle. The fissile nuclide ^{233}U, produced in thorium fuel cycles, is categorised under the same basis as plutonium.

There are four key physical characteristics to take into account when assessing the difficulty of using fissile materials for military purposes:

- The critical mass of a bare homogeneous sphere of fissile material, that is without a neutron reflector.

- The spontaneous neutron emission, which should be as low as possible to guard against the effect of pre-detonation. For highly neutron-emitting fissile materials, it becomes practically impossible to design a "reliable" weapon.

- The intrinsic radiation heat generated within the fissile material itself. Excessive heat can complicate the process of making a weapon or jeopardise its operation.

- The levels of external radiation associated with the fissile material must be considered because of the impact on radiological protection of the personnel handling these materials, as well as for reasons of potential radiation damage to electronic components.

The following remarks can be made based on these criteria:

- The mass of ^{233}U needed to make an atomic weapon is not altogether different from the mass required to make a plutonium weapon, as the critical masses of ^{233}U and plutonium are fairly close.

- The most important difference between uranium (either ^{233}U or ^{235}U) and plutonium originates from spontaneous neutron emission, which in plutonium is mainly due to the isotope ^{240}Pu – present in more or less significant proportions depending on the origin of the plutonium – and in particular the burn-up of the fuel from which plutonium was extracted.

- With ^{233}U, it is possible, due to its very low spontaneous emission of neutrons, to produce a simpler type of weapon than with plutonium, which requires the use of a much more complex implosion device. This simplification in the case of ^{233}U can, however, be mitigated by the high alpha activity of ^{233}U, which could trigger (alpha, n) reactions on light elements that may be present in trace amounts in the fissile material, causing an unwanted emission of neutrons. Nevertheless, this process would produce far fewer neutrons than the spontaneous emission of neutrons from ^{240}Pu, and its effects can be minimised by reducing the levels of contamination in light elements.

- Heat emission from ^{233}U is generally lower than plutonium and does not pose a major problem.

It is possible to make an atomic weapon with ^{233}U, and such devices have in the past been fabricated and detonated.

Self-protection from ^{232}U

The presence of ^{233}U (specifically in the case of thorium-based fuel cycles) is often cited as providing self-protection against proliferation. This is primarily due to the decay products of ^{233}U, which are strong gamma emitters. ^{208}Tl, for example, upon decay emits a 2.6 MeV gamma ray, leading to significant shielding and remote handling needs for any reprocessing activity.

However, the degree of proliferation resistance that this gamma field provides is in reality heavily dependent upon the threat scenario, in other words on the facilities available to the proliferators, their financial capacity to build or otherwise acquire shielded facilities, and conversely, their willingness to expose themselves or personnel handling these materials to radiation doses.

Economic Aspects

Given that the thorium cycle has never been deployed at an industrial scale, there is no accurate data on the costs associated with the different stages of this cycle. There is no supply chain of thorium and consequently no indexed thorium market. It is therefore difficult to provide meaningful cost projections for any system using thorium, but it is nevertheless possible to draw some general conclusions by estimating costs as compared to conventional uranium-plutonium fuel cycles at different stages.

The case of a thorium fuel cycle can be qualitatively compared to the French reference case, where a closed uranium-plutonium fuel cycle is implemented on a large industrial scale.

Cost breakdown of the nuclear fuel cycle	
Uranium	24.6%
Uranium conservation	3.3%
Uranium enrichment	21.3%
Fuel fabrication	16.4%
Interim storage and processing of spent fuel+recycling	26.2%
Final disposal of ultimate waste	8.2%

Considering the different aspects of the above breakdown, the following remarks can be made.

In terms of the raw material, the comparison of uranium with thorium cannot be based on a market price, since such a market price does not currently exist for thorium. It could be necessary to seek new resources once available stocks of thorium are exhausted. However, as thorium would be extracted together with other marketable materials (such as rare earth elements or titanium, for instance), the price would probably be much lower than that of uranium, especially as exploitable deposits are mostly open air, which facilitates the recovery of minerals.

The enrichment step is not applicable for a closed thorium cycle in equilibrium, but must still be considered for the transition phase if the fissile material used initially is in the form of LEU. In such cases, in particular for LEU at close to 20% enrichment, it would generally be necessary to feed the cycle with significantly larger amounts of natural uranium and separative work units (SWU) than for the standard uranium cycle for the same energy output, which would be a significant drawback in terms of the general optimisation of uranium consumption. The amount of this surplus largely depends on, for example, reactor type, and fuel burn-up and fuel management schemes. For thorium based cycles using recycled fissile material (plutonium or ^{233}U), the cost associated with these materials heavily depends on the back end of the chosen fuel cycle.

When considering the fabrication or re-fabrication of fresh or recycled thorium-based fuels, it is necessary to distinguish the type of fissile material chosen:

- If LEU is used as fissile material in once-through cycles, two different kinds of nuclear materials (thorium and LEU) would have to be managed, which leads to additional costs compared to the manufacture of standard enriched uranium fuels? This is particularly the case for the different types of heterogeneous fuel management options being considered in different reactor types, where thorium and uranium are physically separated. Specific studies related to homogeneous thorium-uranium fuel cycles in light water reactors have shown that these fuel cycles are not economically competitive over conventional uranium cycles with current fuel management strategies.

- If plutonium is used with thorium in once-through cycles, processes should not be very different from those that are used today in the manufacture of MOX fuel, and therefore the costs should be comparable.

- If the fissile material is ^{233}U, which implies closing the fuel cycle, the presence of the radioactive daughter products of ^{233}U would require operation behind radiological protection. Recycle fuel fabrication would certainly generate very significant additional costs, which are difficult to estimate since there are no technically and economically proven processes and equipment that have been developed and demonstrated for remotely-operated recycle fuel fabrication within a fully shielded and contained facility, especially at the large industrial scale that would be needed. Essentially, no development has been done at the pilot scale with uranium containing several thousand ppm of ^{233}U.

The final step in the back end of a closed nuclear cycle is the final disposal of medium- and high-level long-lived radioactive waste, for which the solution adopted almost universally is that of a deep geological repository. There is no reason to believe that there is a significant economic difference between uranium and thorium cycles in this final stage, especially as it is assumed here that only residual waste from reprocessing is disposed of after separation of all recyclable materials.

Special Nuclear Material

"Special nuclear material" (SNM) is defined by Title I of the Atomic Energy Act of 1954 as plutonium, uranium-233, or uranium enriched in the isotopes uranium-233 or uranium-235, but does not include source material. The definition includes any other material that the Commission determines to be special nuclear material. The NRC has not declared any other material as special nuclear material.

Where does Special Nuclear Material Come From?

Uranium-233 and plutonium do not occur naturally but are produced by the irradiation of source material or special nuclear material in nuclear reactors and could be extracted from used fuel or targets by chemical separation. Extracting SNM from used fuel is called reprocessing. Plutonium is produced in reactors that use uranium as fuel or targets. Uranium-233 is produced in reactors that use thorium as fuel or targets. No U.S. commercial reprocessing plant is currently licensed by the NRC for operation. Uranium enriched in uranium-235 is created by an enrichment facility.

Why is Control of Special Nuclear Material Important?

Congress enacted Title I of the Atomic Energy Act of 1954, as part of President Eisenhower's Atoms for Peace program, including the clause:

- Source and special nuclear material, production facilities, and utilization facilities are affected with the public interest, and regulation by the United States of the production and utilization of atomic energy and of the facilities used in connection therewith is necessary in the national interest to assure the common defense and security and to protect the health and safety of the public.

- Special nuclear material is only mildly radioactive, but it includes fissile isotopes — uranium-233, uranium-235, and plutonium-239 — that, in concentrated form, could be used as the primary ingredients of nuclear explosives. These materials, in amounts greater than formula quantities, are defined as "strategic special nuclear material" (SSNM). The uranium-235 content of low-enriched uranium can be concentrated (i.e., enriched) to make highly enriched uranium, the primary ingredient of some nuclear explosive designs. To ensure the safety of the public and the security of the Nation, the NRC requires facilities licensed to possess SNM to establish and maintain security programs that protect the material and prevent the loss or theft of SNM.

References

- Nuclear-materials, nuclear-weapons: nuclearfiles.org, Retrieved 23 August, 2019
- What-is-uranium-how-does-it-work, nuclear-fuel-cycle: world-nuclear.org, Retrieved 02 June, 2019

- Plutonium-factsheet: ieer.org, Retrieved 25 April, 2019

- Sp-nucmaterials: nrc.gov, Retrieved 03 March, 2019

- Ade, B. et al. (2014), Safety and Regulatory Issues of the Thorium Fuel Cycle, NUREG/CR-7176, US NRC Publications, Washington, DC

4

Nuclear Reactor: Types and Design

A device which is used for initiation and maintenance of a self-sustained nuclear chain reaction is referred to as a nuclear reactor. It is used in generating electricity and nuclear marine propulsion. This chapter delves into the various types of nuclear reactor and their designs to provide an extensive understanding of this subject.

A nuclear reactor is a system that contains and controls sustained nuclear chain reactions. Reactors are used for generating electricity, moving aircraft carriers and submarines, producing medical isotopes for imaging and cancer treatment, and for conducting research.

Fuel, made up of heavy atoms that split when they absorb neutrons, is placed into the reactor vessel (basically a large tank) along with a small neutron source. The neutrons start a chain reaction where each atom that splits releases more neutrons that cause other atoms to split. Each time an atom splits, it releases large amounts of energy in the form of heat. The heat is carried out of the reactor by coolant, which is most commonly just plain water. The coolant heats up and goes off to a turbine to spin a generator or drive shaft. Nuclear reactors are just exotic heat sources.

Main Components

- The core of the reactor contains all of the nuclear fuel and generates all of the heat. It contains low-enriched uranium (<5% U-235), control systems, and structural materials. The core can contain hundreds of thousands of individual fuel pins.

- The coolant is the material that passes through the core, transferring the heat from the fuel to a turbine. It could be water, heavy-water, liquid sodium, helium, or something else. In the US fleet of power reactors, water is the standard.

- The turbine transfers the heat from the coolant to electricity, just like in a fossil-fuel plant.

- The containment is the structure that separates the reactor from the environment. These are usually dome-shaped, made of high-density, steel-reinforced concrete. Chernobyl did not have a containment to speak of.

- Cooling towers are needed by some plants to dump the excess heat that cannot be converted to energy due to the laws of thermodynamics. These are the hyperbolic icons of nuclear energy. They emit only clean water vapor.

Fuel Pins

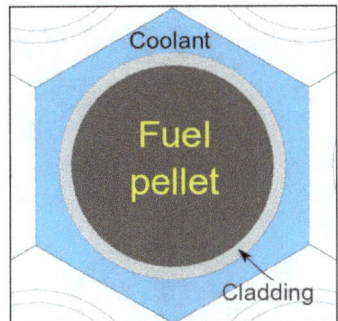

The smallest unit of the reactor is the fuel pin. These are typically uranium-oxide (UO_2), but can take on other forms, including thorium-bearing material. They are often surrounded by a metal tube (called the cladding) to keep fission products from escaping into the coolant.

Fuel Assembly

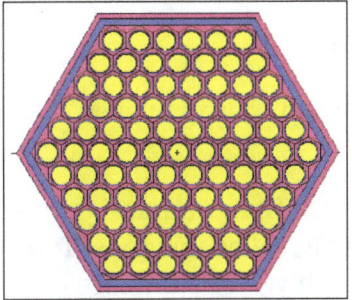

Fuel assemblies are bundles of fuel pins. Fuel is put in and taken out of the reactor in assemblies. The assemblies have some structural material to keep the pins close but not touching, so that there's room for coolant.

Full Core

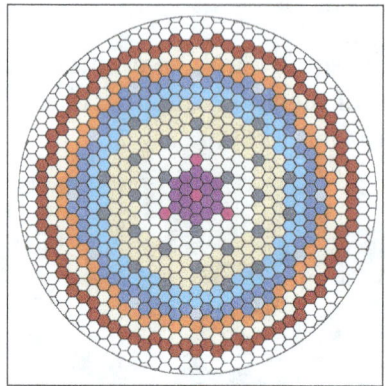

This is a full core, made up of several hundred assemblies. Some assemblies are control assemblies. Various fuel assemblies around the core have different fuel in them. They vary in enrichment and age, among other parameters. The assemblies may also vary with height, with different enrichments at the top of the core from those at the bottom.

Nuclear Reactor Work

Nuclear reactors are the heart of a nuclear power plant.

They contain and control nuclear chain reactions that produce heat through a physical process called fission. That heat is used to make steam that spins a turbine to create electricity.

With more than 450 commercial reactors worldwide, including 96 in the United States, nuclear power continues to be one of the largest sources of reliable carbon-free electricity available.

Nuclear Fission Creates Heat

The main job of a reactor is to house and control nuclear fission—a process where atoms split and release energy.

Reactors use uranium for nuclear fuel. The uranium is processed into small ceramic pellets and stacked together into sealed metal tubes called fuel rods. Typically more than 200 of these rods are bundled together to form a fuel assembly. A reactor core is typically made up of a couple hundred assemblies, depending on power level.

Inside the reactor vessel, the fuel rods are immersed in water which acts as both a coolant and moderator. The moderator helps slow down the neutrons produced by fission to sustain the chain reaction.

Control rods can then be inserted into the reactor core to reduce the reaction rate or withdrawn to increase it.

The heat created by fission turns the water into steam, which spins a turbine to produce carbon-free electricity.

All commercial nuclear reactors in the United States are light-water reactors. This means they use normal water as both a coolant and neutron moderator.

There are two types of light-water reactors operating in America.

Pressurized Water Reactors

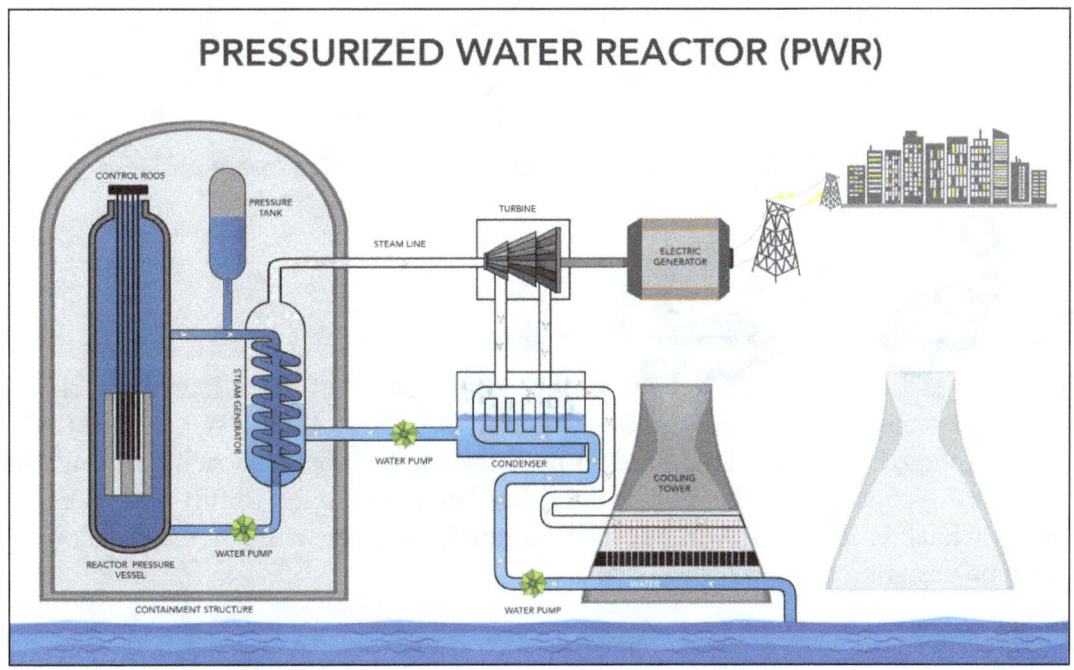

More than 65% of the commercial reactors in the United States are pressurized-water reactors or PWRs. These reactors pump water into the reactor core under high pressure to prevent the water from boiling.

The water in the core is heated by nuclear fission and then pumped into tubes inside a heat exchanger. Those tubes heat a separate water source to create steam. The steam then turns an electric generator to produce electricity.

The core water cycles back to the reactor to be reheated and the process is repeated.

Boiling Water Reactors

Roughly a third of the reactors operating in the United States are boiling water reactors (BWRs).

BWRs heat water and produce steam directly inside the reactor vessel. Water is pumped up through the reactor core and heated by fission. Pipes then feed the steam directly to a turbine to produce electricity.

The unused steam is then condensed back to water and reused in the heating process.

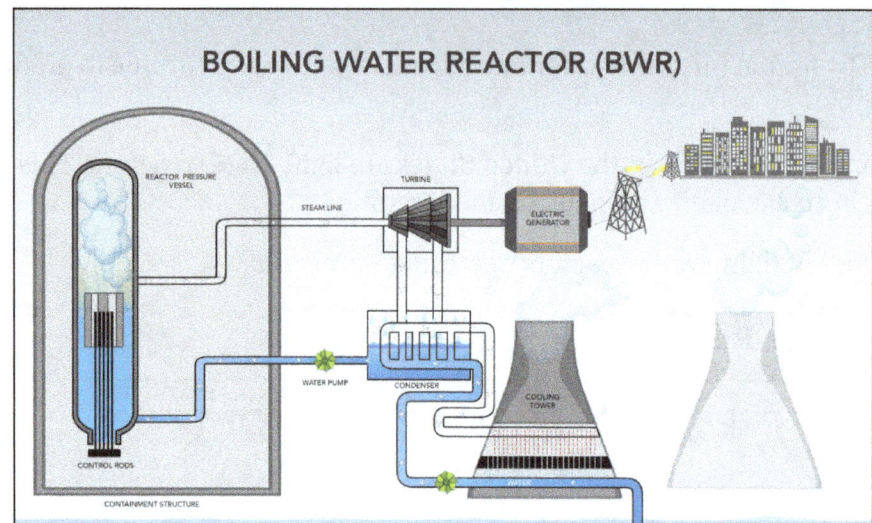

Nuclear Reactor Types

Many different reactor systems have been proposed and some of these have been developed to prototype and commercial scale. Six types of reactor (Magnox, AGR, PWR, BWR, CANDU and RBMK) have emerged as the designs used to produce commercial electricity around the world. A further reactor type, the so-called fast reactor, has been developed to full-scale demonstration stage. These various reactor types will now be described, together with current developments and some prototype designs.

Gas Cooled and Graphite Moderated

Of the six main commercial reactor types, two (Magnox and AGR) owe much to the very earliest reactor designs in that they are graphite moderated and gas cooled. Magnox reactors were built in the UK from 1956 to 1971 but have now been superseded. The Magnox reactor is named after the magnesium alloy used to encase the fuel, which is natural uranium metal. Fuel elements consisting of fuel rods encased in Magnox cans are loaded into vertical channels in a core constructed of graphite blocks. Further vertical channels contain control rods (strong neutron absorbers) which can be inserted or withdrawn from the core to adjust the rate of the fission process and, therefore, the heat output. The whole assembly is cooled by blowing carbon dioxide gas past the fuel cans, which are specially designed to enhance heat transfer. The hot gas then converts water to steam in a steam generator. Early designs used a steel pressure vessel, which was surrounded by a thick concrete radiation shield. In later designs, a dual-purpose concrete pressure vessel and radiation shield was used.

In order to improve the cost effectiveness of this type of reactor, it was necessary to go to higher temperatures to achieve higher thermal efficiencies and higher power densities to reduce capital costs. This entailed increases in cooling gas pressure and changing from Magnox to stainless steel cladding and from uranium metal to uranium dioxide fuel. This in turn led to the need for an increase in the proportion of U235 in the fuel. The resulting design, known as the Advanced Gas-Cooled Reactor, or AGR. still uses graphite as the moderator and, as in the later Magnox designs, the steam generators and gas circulators are placed within a combined concrete pressure-vessel/radiation shield.

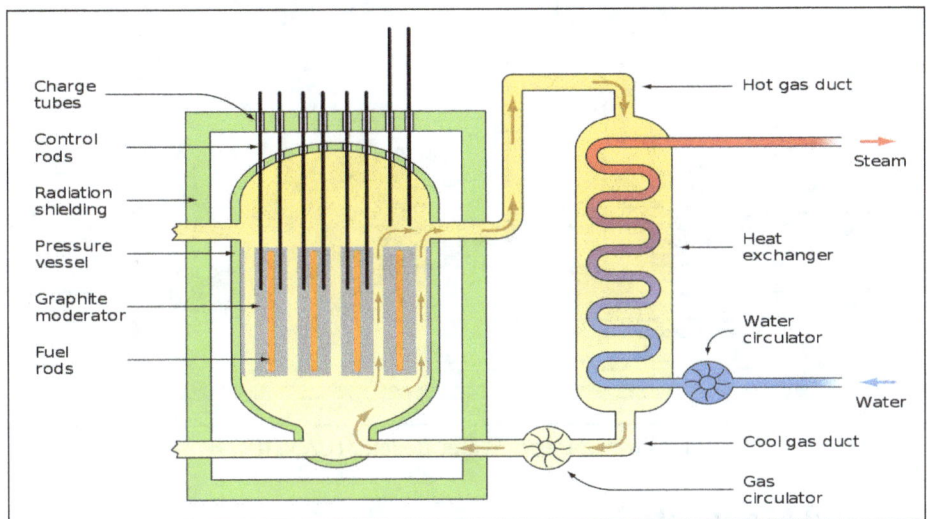

Basic Gas-Cooled Reactor (MAGNOX).

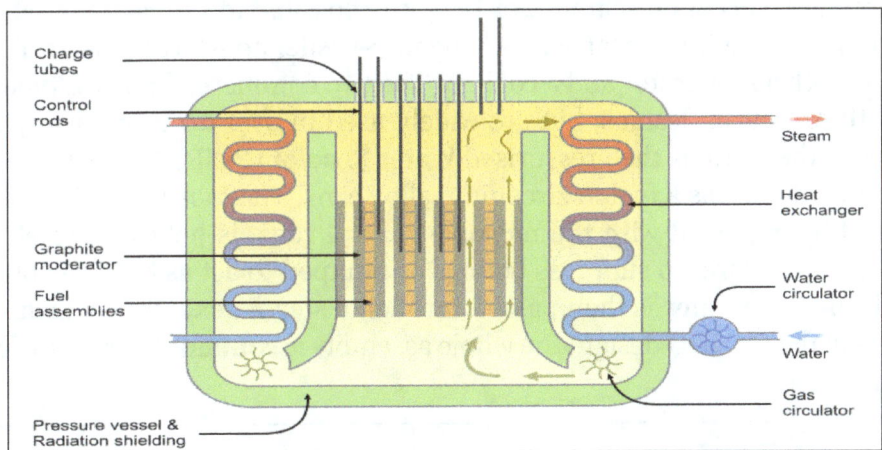

Advanced Gas-Cooled Reactor (AGR).

Heavy Water Cooled and Moderated

The only design of heavy water moderated reactor in commercial use is the CANDU, designed in Canada and subsequently exported to several countries. In the CANDU reactor, enriched uranium dioxide is held in zirconium alloy cans loaded into horizontal zirconium alloy tubes. The fuel is cooled by pumping heavy water through the tubes (under high pressure to prevent boiling) and then to a steam generator to raise steam from ordinary water (also known as natural or light water) in the normal way. The necessary additional moderation is achieved by immersing the zirconium alloy tubes in an unpressurised container (called a callandria) containing more heavy water. Control is effected by inserting or withdrawing cadmium rods from the callandria. The whole assembly is contained inside the concrete shield and containment vessel.

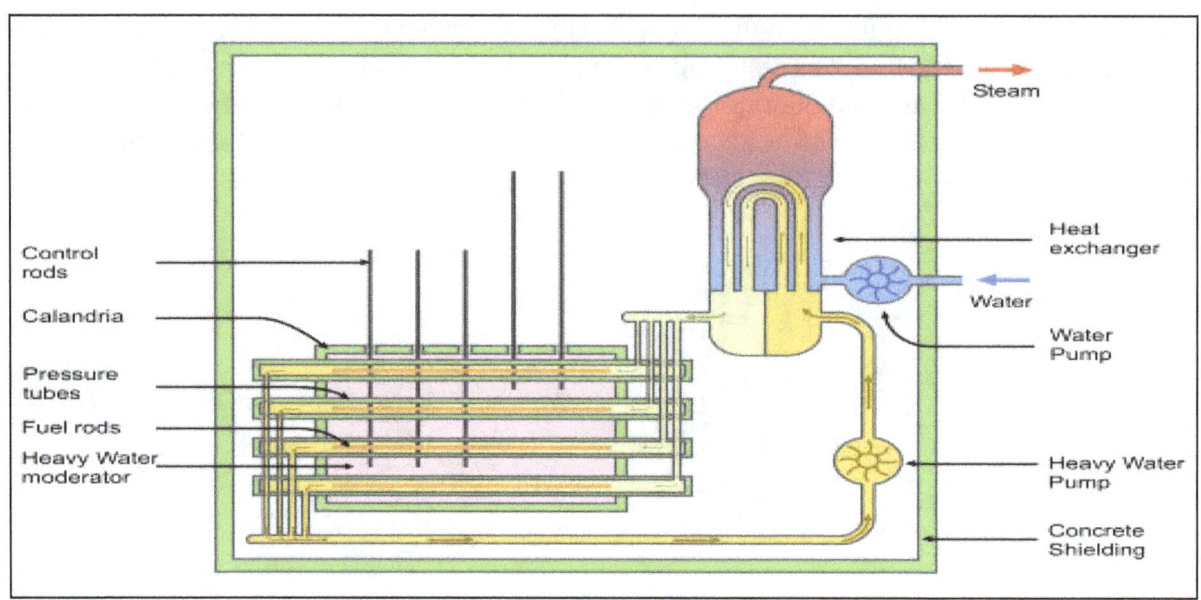

Pressurised Heavy Water Reactor (CANDU).

Water Cooled and Moderated

By moving to greater levels of enrichment of U^{235}, it is possible to tolerate a greater level of neutron absorption in the core (that is, absorption by non-fissile, non-fertile materials) and thus use ordinary water as both a moderator and a coolant. The two commercial reactor types based on this principle are both American designs, but are widely used in over 20 countries. The most widely used reactor type in the world is the Pressurised Water Reactor (PWR) which uses enriched (about 3.2% U^{235}) uranium dioxide as a fuel in zirconium alloy cans. The fuel, which is arranged in arrays of fuel "pins" and interspersed with the movable control rods, is held in a steel vessel through which water at high pressure (to suppress boiling) is pumped to act as both a coolant and a moderator. The high-pressure water is then passed through a steam generator, which raises steam in the usual way. As in the CANDU design, the whole assembly is contained inside the concrete shield and containment vessel.

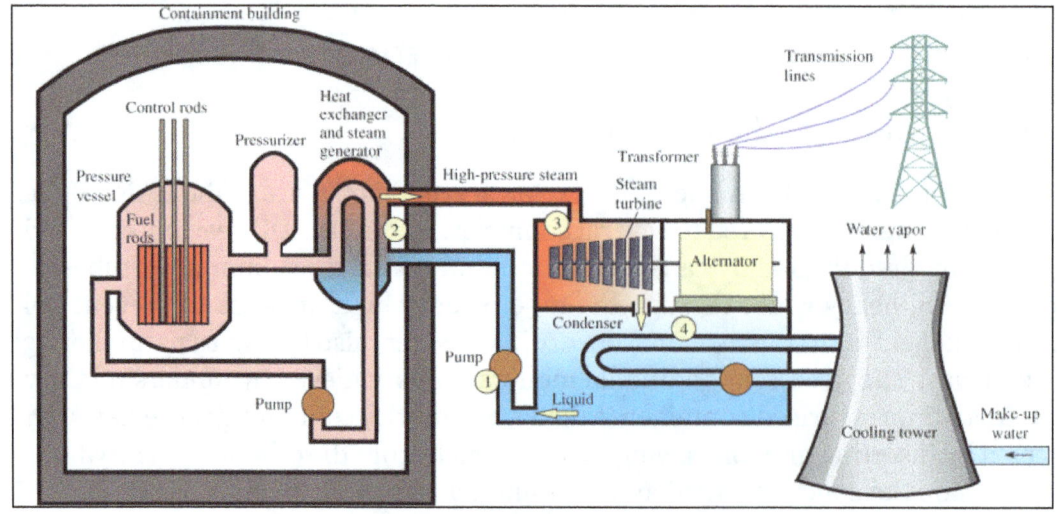

Pressurised Water Reactor (PWR).

The second type of water cooled and moderated reactor does away with the steam generator and, by allowing the water within the reactor circuit to boil, it raises steam directly for electrical power generation. This, however, leads to some radioactive contamination of the steam circuit and turbine, which then requires shielding of these components in addition to that surrounding the reactor. Such reactors, known as Boiling Water Reactors (BWRs), are in use in some ten countries throughout the world.

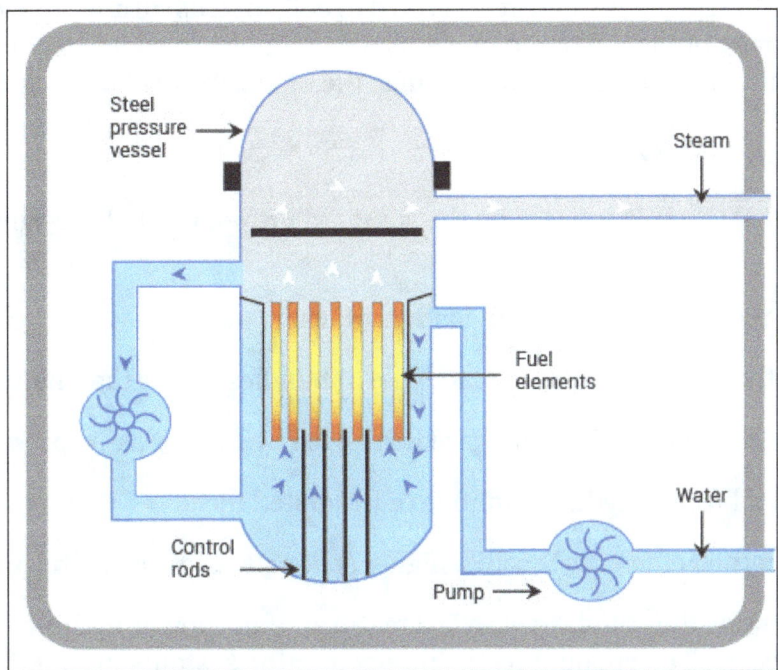

Boiling Water Reactor (BWR).

Water Cooled and Graphite Moderated

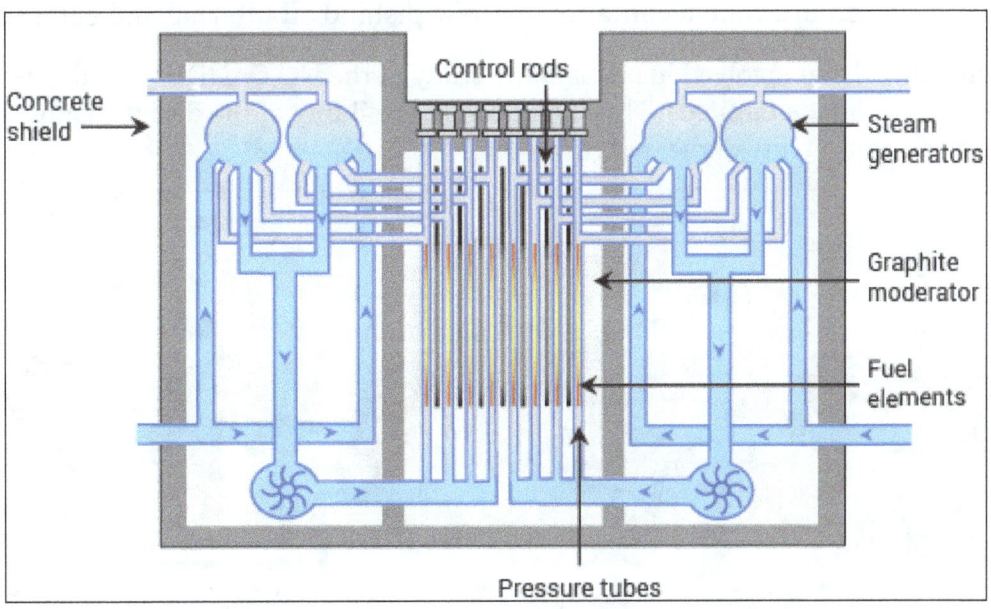

RBMK REACTOR Boiling Light Water, Graphite Moderated Reactor.

At about the same time as the British gas cooled, graphite moderated Magnox design was being commissioned at Calder Hall in 1956, the Russians were testing a water cooled, graphite moderated plant at Obninsk. The design, known as the RBMK Reactor, has been developed and enlarged, and many reactors of this type have been constructed in the USSR, including the illfated Chernobyl plant. The layout consists of a large graphite core containing some 1700 vertical channels, each containing enriched uranium dioxide fuel (1.8% U^{235}). Heat is removed from the fuel by pumping water under pressure up through the channels where it is allowed to boil, to steam drums, thence driving electrical turbo-generators. Many of the major components, including pumps and steam drums, are located within a concrete shield to protect operators against the radioactivity of the steam.

Next-Generation CANDU

Next-Generation (NG) CANDU is based on the standard proven CANDU design. It introduces new features:

- Light water reactor coolant system instead of heavy water.

- Use of slightly enriched uranium oxide fuel in bundles rather than natural uranium fuel.

- Compact reactor core design: core size is reduced by half for same power output.

- Extended fuel life with reduced volume of irradiated fuel.

- Improved thermal efficiency through higher steam pressure steam turbines.

The NG CANDU retains the standard CANDU features of on-power fuelling, simple fuel design and flexible fuel cycles. The steam and turbine generator systems are similar to those in advanced pressurised water reactor systems. For safety, NG CANDU design includes two totally independent safety shutdown systems and an inherent passive emergency fuel cooling capability in which the moderator absorbs excess heat. The whole of the primary system and the steam generators are housed in a robust containment to withstand all internal and external events.

British Energy have been involved in a feasibility study of the NG CANDU with the vendor AECL (Atomic Energy of Canada Limited). This included the feasibility of the design against UK criteria and in particular licensability of the design.

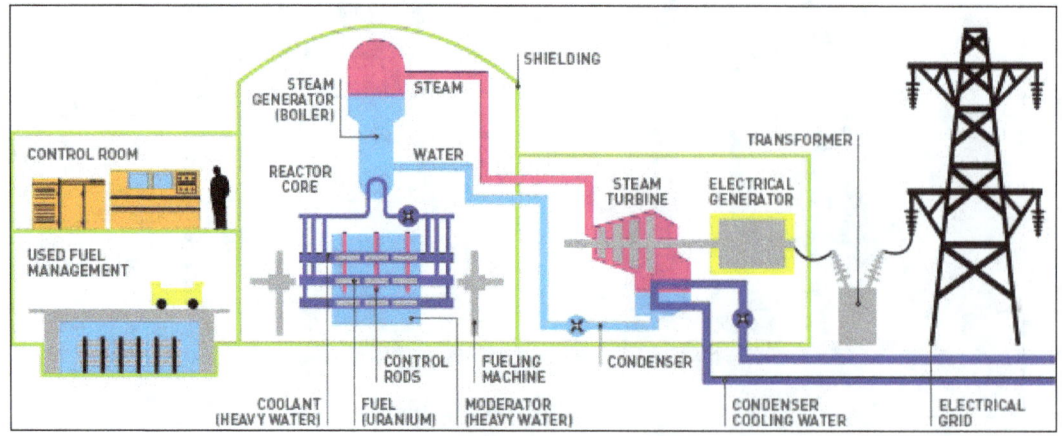

NG CANDU Flow Diagram.

Advanced Pressurised Water Reactor AP1000

As part of a co-operative programme with the US Department of Energy, the Westinghouse Company, which is owned by BNFL, have developed an Advanced PWR with predominantly passive safety systems. Termed the AP600 (600MWe) it is the most up-to-date design licensed in the United States. BNFL have also developed the AP1000 (1000MWe) with similar safety features to the smaller version but gaining in economies of scale. The advanced passive design is a development of the PWR design at Sizewell B.

Key design features of the AP designs are:

- Simplification of standard PWR designs with less piping, fewer valves, less control cabling and reduced seismic building volumes.

- Modular manufacturing techniques giving a shorter construction schedule (for the AP600 plant 36 months from first concrete to fuel loading).

- Passive safety systems using only natural forces such as gravity, natural circulation and compressed gas. Fans, pumps, diesels and chillers are not required for safety, nor is operator intervention. A few simple valves are used to align the passive safety systems when required, in most instances the valves are 'fail safe' in that on loss of power they move to the safety position.

- The passive cooling systems include core cooling, providing residual heat removal, reactor coolant make-up and safety-injection, and containment cooling which provides the safety related ultimate heat sink for the plant.

- Operating lifetime of 60 years with a design plant availability of 90%+.

- Probabilistic risk assessment has been used as an integral part of the design process with numerous fine design changes being made as a result of the PRA studies. The net effect of the overall design approach is that the predicted core damage frequency is about a factor of 100 better than current plant designs.

British Energy have been involved in a feasibility study with BNFL/Westinghouse covering:

- The feasibility of the design against UK criteria and, in particular, the licensability of the design.

- The technical suitability of AP1000 reactors on existing sites.

- The economic case for the plant and potential funding models.

Prototype Designs

Designs now being considered for the longer term include:

International Reactor, Innovative and Secure

This is based on a small LWR concept with secure safety aspects built in. The design will be modular and flexible and achieve economic competitiveness.

Pebble Bed Modular Reactor

The reactor is a helium-cooled graphite moderated unit of 100MWe which drives a gas turbine linked to a generator giving up to 50% efficiency. Key design features:

- Fuel elements are spherical 'pebbles' 60mm in diameter of graphite containing tiny spheres of uranium dioxide coated with carbon and silicon carbide. This coating retains the gaseous and volatile fission products generated in operation.

- The reactor consists of a vertical steel pressure vessel, 6m in diameter and about 20m high. It is lined with graphite bricks drilled with vertical holes to house the control rods.

- Helium is used as the coolant and transfers heat to a closed cycle gas turbine and generator.

- When fully loaded the core contains 310,000 fuel spheres; re-fuelling is done on-line with irradiated spheres being withdrawn at the base of the reactor and fresh fuel elements being added at the top.

- The PBMR has inherent passive safety features that require no operator intervention. Removal of decay heat is achieved by radiation, conduction and convection. The combination of very low power density of the core and temperature resistance of the fuel in millions of independent particles underpins the safety assurance of the design.

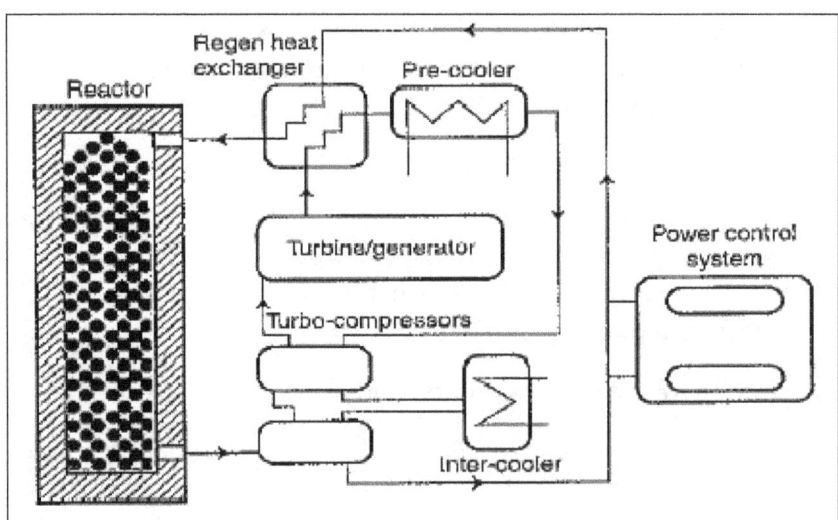

Circuit of Pebble Bed Modular Reactor.

The PBMR design takes forward the approach originally developed in Germany (AVR 15MW experimental pebble bed reactor and Thorium High-Temperature Reactor THTR 300MWe) and is being developed by Eskom, the South African electrical utility, for application in South Africa initially through a demonstration plant. Exelon (the major US utility) and BNFL are supporting this venture to develop and commercialise the PBMR.

Fast Reactors All of today's commercially successful reactor systems are "thermal" reactors, using slow or thermal neutrons to maintain the fission chain reaction in the U235 fuel. Even with the enrichment levels used in the fuel for such reactors, however, by far the largest numbers of atoms present are U238, which are not fissile.

Consequently, when these atoms absorb an extra neutron, their nuclei do not split but are converted into another element, Plutonium. Plutonium is fissile and some of it is consumed in situ, while some remains in the spent fuel together with unused U^{235}. These fissile components can be separated from the fission product wastes and recycled to reduce the consumption of uranium in thermal reactors by up to 40%, although clearly thermal reactors still require a substantial net feed of natural uranium.

It is possible, however, to design a reactor which overall produces more fissile material in the form of Plutonium than it consumes. This is the fast reactor in which the neutrons are unmoderated, hence the term "fast". The physics of this type of reactor dictates a core with a high fissile concentration, typically around 20%, and made of Plutonium. In order to make it breed, the active core is surrounded by material (largely U^{238}) left over from the thermal reactor enrichment process. This material is referred to as fertile, because it converts to fissile material when irradiated during operation of the reactor.

Due to the absence of a moderator, and the high fissile content of the core, heat removal requires the use of a high conductivity coolant, such as liquid sodium. Sodium circulated through the core heats a secondary loop of sodium coolant, which then heats water in a steam generator to raise steam. Otherwise, design practice follows established lines, with fuel assemblies clad in cans and arranged together in the core, interspersed with movable control rods. The core is either immersed in a pool of coolant, or coolant is pumped through the core and thence to a heat exchanger. The reactor is largely unpressurised since sodium does not boil at the temperatures experienced, and is contained within steel and concrete shields.

The successful development of fast reactors has considerable appeal in principle. This is because they have the potential to increase the energy available from a given quantity of uranium by a factor of fifty or more, and can utilise the existing stocks of depleted uranium, which would otherwise have no value.

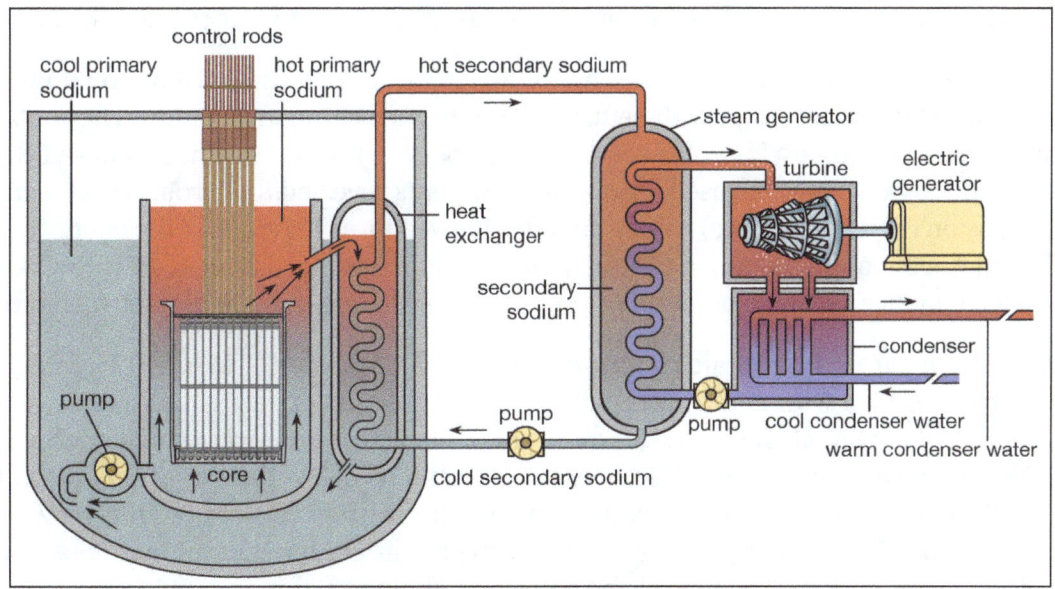

Sodium-cooled Fast Reactor.

Fast reactors, however, are still currently at the prototype or demonstration stage. They would be more expensive to build than other types of nuclear power station and will therefore become commercial only if uranium or other energy prices substantially increase.

Fusion

All the reactors outlined before are fission reactors. Energy can also be produced by fusing together the nuclei of light elements. This is the process which provides the energy source in the sun and other stars. The idea of releasing large amounts of energy by the controlled fusion of the nuclei of atoms such as deuterium and tritium is very attractive because deuterium occurs naturally in seawater.

Unfortunately, controlled fusion has turned out to be an extraordinarily difficult process to achieve. For the reaction to proceed, temperatures in excess of one hundred million degrees must be obtained and high densities of deuterium and tritium must be achieved and retained for a sufficient length of time. So far, it has not proved possible to sustain these requirements simultaneously in a controlled way. Even if this can be achieved eventually, the process must be capable of being developed in a form which will allow power to be generated cost effectively and continuously over a long period. It is very unlikely that this could be achieved until well into the twenty-first century.

Design of a Nuclear Reactor

Design considerations have engaged human minds since time immemorial. Consider the example of fire. The control and use of fire marks a dividing line between man and other mammals. The flintstones used to ignite a spark underwent a number of changes. The variety of arrangements of logs, twigs and dried leaves leading to a sustained combustion would make for a fascinating study. Some of these arrangements have been handed down to us from prehistoric times and can still be seen in rural kitchens. The present topic is also concerned with design. It employs pre-college physics to arrive at an understanding of a system as complex as a nuclear reactor (NR).

Around 1938, the German physicist Otto Hahn discovered that uranium nucleus could be split into two smaller nuclei by bombarding it with neutrons, later termed as 'neutron induced fission'1. A back-of-the-envelope calculation by Lise Meitner and her nephew Otto Robert Frisch showed that in fission, about 200 MeV energy is released. Further experiments revealed that around 2–3 neutrons are emitted per fission (fission neutrons) having energies between 1–2 MeV; lower the energy of the neutrons inducing fission, more the probability of fission. To harness this fission energy, the fission neutrons are used to induce fission in other uranium nuclei to sustain a chain of fissions (chain reaction).

Naturally occurring uranium contains 0.72% of ^{235}U and 99.28% of ^{238}U. While ^{235}U, categorised as fissile, undergoes fission with neutrons of any energy, ^{238}U, categorised as fissionable, undergoes fission with high energy neutrons only. But due to high inelastic scattering of high energy neutrons from ^{238}U the energy of most of the fission neutrons is reduced below the threshold energy to cause fission in ^{238}U. So, sustaining a chain reaction even in 99.28% of ^{238}U is not possible. Now to sustain fission in 0.72% of ^{235}U, the energy of the fission neutrons should be reduced to increase the probability of fission. Enrico Fermi achieved the sustained chain reaction in 0.72% of ^{235}U in the first man-made NR named Chicago Pile-1 in 1942 by making the fission neutrons collide with low-mass number element called moderators (graphite) to lose energy. For efficient neutron utilisation, they should have a low affinity to absorb neutrons. Later, in the year 1954, USSR's Obninsk Nuclear Power Plant became the world's first to generate around 5 MW of electric power.

Reactors are categorized broadly into two types: thermal and fast reactors. Fast reactor technology is at the developmental stage and almost all commercial reactors are thermal.

Thermal Reactors

In the thermal reactors, fission is caused by neutrons with their energy thermalised to the temperature of the moderator. The fuel for these reactors is UO_2. There are several types of thermal reactors from the compact to the very large reactors and with a variety of technologies. We mention the two most popular types of thermal reactors.

Boiling Water Reactors

The most successful and compact reactors are the boiling water reactors. They are called pressure vessel reactors where the moderator is light (normal) water, which also acts as coolant. The water surrounds the fuel bundles just like in the case of a heater rod immersed in water. The heat from the fuel bundles is transferred directly to the water which is allowed to boil. This same boiling water acts as the moderator for the fission neutrons. The steam generated from boiling is collected and sent to turbines. For this reason, they are most efficient and the use of light water makes them most compact. But since the reactor has to be vacuum sealed for the collection of steam, refueling is one of the major issues. After a sufficient duration of operation, the reactor requires to be shut down for refueling. Further, absorption of neutrons in light water is high. In order to compensate for the loss of neutrons, the fuel UO_2 requires enrichment of ^{235}U, sometimes as much as 20%, and this is a difficult technology.

Pressurised Heavy Water Reactors

The second most popular are the pressure tube reactors, where coolant and the moderator are physically separated, though they both are heavy water. While the coolant is used to remove heat from the fuel, the moderator is used to reduce the energy of the high energy neutrons. The name 'pressure tube reactors' is because the coolant in these reactors is pressurised and the tubes through which the coolant flows has to withstand this pressure. Since the moderator and coolant are kept separate, the design of these reactors for the coolant flow is not simple. They are nevertheless very popular since they can use natural uranium without any enrichment.

Fast Reactors

In fast reactors fission is sustained by high energy (1Mev) neutrons inducing fission. So the fuel is different from that of thermal reactors.The fuels ^{233}U and ^{239}Pu for fast reactors are not available in Nature but created respectively when ^{232}Th and ^{238}U transmutes after absorbing a neutron. The reactors in which the fuel for fast reactors are produced are called fast breeder reactors. Fast reactors do not have moderators and coolant has high mass number. When high energy neutrons induce fission, the fission neutron emission increases hence ^{233}U or ^{239}Pu can be bred by placing ^{232}Th or UO_2 inside the reactor. The coolant in this reactor is liquid sodium, a material difficult to handle. Also, the neutron population changes at a faster rate which requires a fast acting system for control. It is these difficulties along with the production of ^{233}U and ^{239}Pu which has delayed the commercial success of fast reactors.

Typical Nuclear Reactor

A typical NR there are fuel bundles which are stacked one above the other to a height H. A fuel bundle contains a cluster of cylindrical fuel pins of solid natural UO_2. Fission neutrons, coming outward from a fuel channel, collide with the moderator, losing energy, and reach the surrounding fuel channels with low enough energy to cause further fissions. Heat generated from fission in the pin is transmitted to the surrounding coolant fluid flowing along its length inside the fuel channel. This cylinder is kept horizontal to avoid pumping the coolant against gravity. The coolant in these type of reactors is again D_2O and is pressurised to avoid boiling. For this reason, such an NR is called Pressurised Heavy Water Reactor (PHWR). The heat from the coolant is removed in a heat exchanger and is used to produce steam which runs the turbines to produce electricity. Here, we shall study some of the physics behind the (i) design of the fuel pin, (ii) role of a moderator and finally (iii) dimensions of a NR of cylindrical geometry. However, the study can be extended to other types of reactors. consists of a cylindrical tank of height H and radius R filled with D_2O (heavy water) called moderator. Cylindrical tubes, called fuel channels are kept axially in a square array. In each fuel channel.

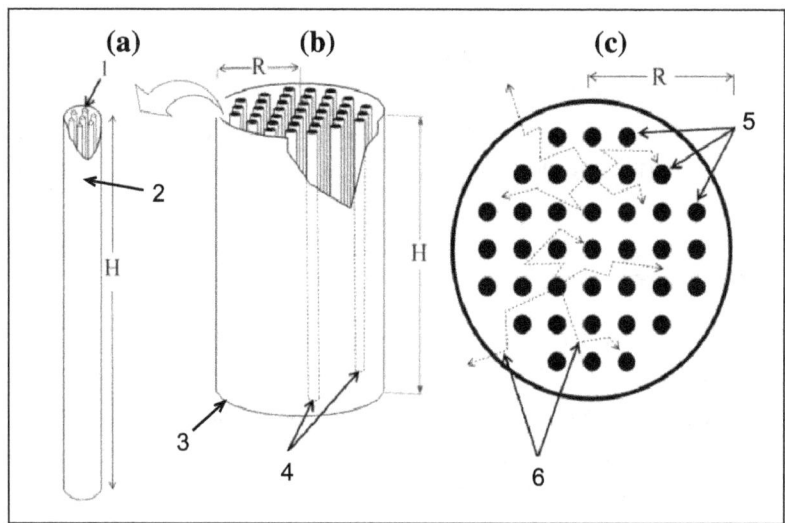

Schematic sketch of the Nuclear Reactor (NR): (a). Enlarged view of a fuel channel. (1: Fuel pins; 2: Pressurised coolant (D_2 O) flows inside channel at an average temperature of 555 K.); (b) A view of the NR. (3: Filled with moderator (D_2 O) maintained at 353 K; 4: Fuel channels.); (c) Top view of NR. (5: Square arrangement of fuel channels; 6: Typical neutron paths.)

Fuel Pin

The fuel pin is the elementary entity of the NR. We shall arrive at the upper limit on the radius of the fuel pin starting from the energy released in a typical nuclear fission. Consider a typical fission reaction of a stationary 235U after it absorbs a neutron of negligible kinetic energy:

$$^{235}U + {}^1n \longrightarrow {}^{94}Zr + {}^{140}Ce + 2\ {}^1n + \Delta E$$

We estimate ΔE (in MeV) – the total fission energy released. Ignoring charge imbalance, this energy released can be calculated using Einstein's famous mass–energy formula ($E = mc^2$),

$$\Delta E = [m(^{235}U)+m(^{1}n)-m(^{94}Zr)-m(^{140}Ce)-2m(^{1}n)]c^2$$

Substituting in terms of unified atomic masses (u) we get $\Delta E = 208.684$ MeV.

Table: Relevant data for UO_2 and nuclear masses.

Nuclear Masses	Data for UO_2
m (^{235}U)	Molecular weight $M_w = 0.270$ kg mol^{-1}
m (^{94}Zr)	Density $\rho = 1.060 \times 10^4$ kg m^{-3}
m (^{140}Ce)	Melting point $T_m = 3.138 \times 10^3$ K
m (^{1}n)	Thermal conductivity $\lambda = 3.280$ W m^{-1} K^{-1}

Let N be the number of ^{235}U atoms per unit volume in natural UO_2. The number of UO_2 molecules per m^3 of the fuel N_1 is given in terms of its density ρ, the Avogadro number N_A and the average molecular weight Mw as:

$$N_1 = \frac{\rho N_A}{M_w} = \frac{10600 \times 6.022 \times 10^{23}}{0.270} = 2.364 \times 10^{28} m^{-3}$$

Each molecule of UO_2 contains one uranium atom. Since only 0.720% of them are ^{235}U, $N = 0.0072 \times N_1 = 1.702 \times 10^{26}$ m^{-3}.

For the study of interaction of neutrons with a target material, we define the following three quantities.

- Neutron flux (φ) is the total number of neutrons crossing a unit area per second.

- The microscopic cross-section σ is the number of (specific) interactions of neutrons per second per unit atom of the target material.

- The macroscopic cross-section Σ is the product of σ and the atom density N of the target material. Let us assume that the neutron flux $\varphi = 2.000 \times 1018$ m^{-2} s^{-1} on the fuel is uniform. The microscopic fission cross-section of a ^{235}U nucleus is $\sigma_f = 5.400 \times 10^{-26}$ m^2.

We will now estimate Q (in Wm^{-3}), the rate of heat production in the pin per unit volume, given that 80.00% of the fission energy is available as heat. Heat energy available per fission $E_f = 0.8 \times 208.7$ MeV $= 2.675 \times 10^{-11}$J. The total cross-section per unit volume is $N \times \sigma f$. Thus, the heat produced per unit volume per unit time,

$$Q = N \times \sigma_f \times \varphi \times E_f = 4.917 \times 10^8 \text{ Wm}^{-3}$$

To estimate the radius of the fuel pin, we consider the steady-state temperature difference between the center (T_c) and the surface (T_s) of the pin which can be expressed as $T_c - T_s = kF(Q, a, \lambda)$, where k is a dimensionless constant and a is the radius of the pin. The functional dependence $F(Q, a, \lambda)$ of the temperature difference can be obtained by solving the Fourier equation for heat conduction. However, there is an appealing method using dimensional analysis. The dimensions of Tc − Ts is temperature. We write this as $T_c - T_s = [K]$. One can similarly write down the dimensions of Q, a and λ. Equating the temperature to powers of Q, a and λ, one could state the following dimensional equation:

$$K = Q^\alpha a^\beta \lambda^\gamma = \left[M L^{-1} T^{-3} \right]^\alpha [L]^\beta \left[M L^1 T^{-3} K^{-1} \right]^\gamma$$

By equating powers of temperature, we get $\gamma = -1$ and $\alpha + \gamma = 0$ by equating powers of mass or time. Therefore, $\alpha = 1$. Next, equating powers of length yields $-\alpha + \beta + \gamma = 0$ which gives $\beta = 2$. Thus we have:

$$T_c - T_s = \frac{Qa^2}{4\lambda}$$

where, $k = 1/4$, and this dimensionless factor can be obtained from the detailed solution of the heat conduction equation. Factors like the efficiency of the heat exchanger, the steam pressure requirement of the turbine, etc., empirically decide the inlet temperature of the coolant as 533 K. To increase the heat transfer ability throughout the cycle, the coolant is pressurised so that there is no boiling. An optimised flow rate of the coolant through the fuel channel increases the temperature by about 44 K making the maximum outlet temperature of the coolant 577 K. We can then estimate the upper limit au on the radius a of the pin. The melting point of UO_2 is 3138 K. This sets a limit on the maximum permissible temperature $(T_c - T_s)$ to be less than $(3138 - 577 = 2561$ K$)$ to avoid 'meltdown'. Thus one may take a maximum of $(T_c - T_s) = 2561$ K. Noting that $\lambda = 3.28$ Wm^{-1} K^{-1} we have:

$$a_u^2 = \frac{2561 \times 4 \times 3.28}{4.917 \times 10^8}$$

This yields $a_u \simeq 8.267 \times 10^{-3}$ m, which constitutes an upper limit on the radius of the fuel pin.

Moderator

Consider the 2D elastic collision between a neutron of mass 1 u and a moderator atom of mass Au. Before collision, all the moderator atoms are considered to be at rest in the laboratory frame (LF). Let $\vec{v_b}$ and $\vec{v_a}$ be the velocities of the neutron before and after collision respectively in the LF. Let $\vec{v_m}$ be the velocity of the center of mass (CM) frame relative to LF and θ be the neutron scattering angle in the CM frame. All the particles involved in collisions are moving at non-relativistic speeds. Let E_b and E_a be the kinetic energies of the neutron, in the LF, before and after the collision respectively. We will estimate E_a/E_b and rewrite it in terms of a dimensionless parameter $\alpha \equiv [(A - 1)/(A + 1)]^2$.

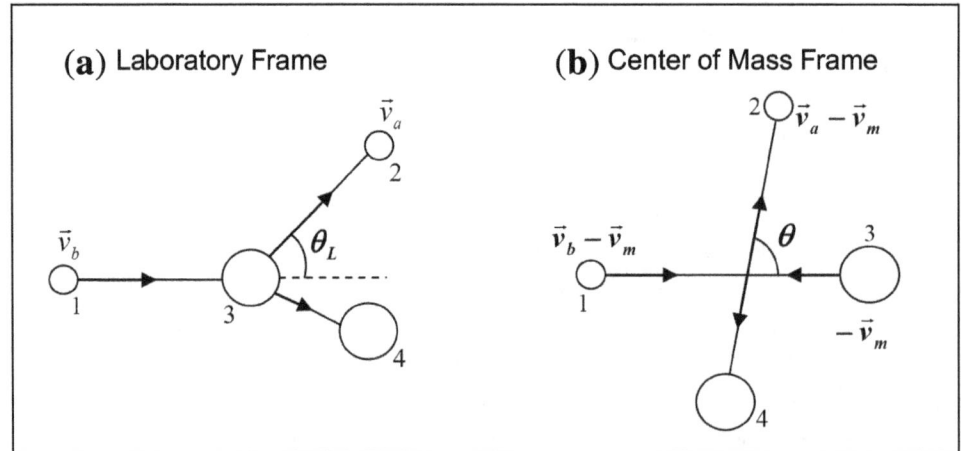

The collision of the neutron labeled 1 with moderator atom labeled 3 in the LF (a) and CM frame (b). The scattered neutron and moderator atom are labeled 2 and 4 respectively.

The collision is shown schematically with θ_L as the scattering angle in LF and θ as the scattering angle in CM frame. From the definition of the CM frame, $v_m = v_b/(A+1)$, before the collision $v_b - v_m = A_{vb}/(A+1)$ and v_m will be the magnitude of the velocities of the neutron and moderator atom respectively. In elastic collision, the particles are scattered in opposite directions in the CM frame and so the speeds remain the same $v = A_{vb}/(A+1)$ and $V = vb/(A+1)$. Since $\vec{v}_a = \vec{v} + \vec{v}_m$, $v_a^2 = v^2 + v^2 m + 2vv_m \cos\theta$. Substituting the values of v and v_m,

$$\frac{v_a^2}{v_b^2} = \frac{E_a}{E_b} = \frac{A^2 + 2A\cos\theta + 1}{(A+1)^2}$$

$$= \frac{1}{2}\left[(1+\alpha) + (1-\alpha)\cos\theta\right]$$

A neutron undergoes a number of collisions in an NR. An average over θ of the above expression, gives the average fraction of energy lost per collision. Since this value will be a function of α only, it is a constant. So on an average, the neutron loses the same fraction of energy per collision. Then the energy E_n after n collisions is:

$$E_n = E_0 e^{-\xi(A)n}$$

where, E_0 is the initial neutron energy and $\xi(A)$ is a constant which is a property of the mass number of the moderator atom. Using the property of logarithms, the above expression can be written as:

$$\xi(A) = \frac{\ln\dfrac{E_0}{E_1} + \ln\dfrac{E_1}{E_2} + \ldots + \ln\dfrac{E_{n-2}}{E_{n-1}} + \ln\dfrac{E_{n-1}}{E_n}}{n} = \overline{\ln\frac{E_b}{E_a}}$$

where, the last step follows for large n and the bar denotes averaging. Assuming scattering is isotropic in the CM frame, the weight function is $\sin\theta$. Thus,

$$\xi(A) = \frac{\displaystyle\int_0^\pi \ln\frac{E_b}{E_a}\sin\theta\,d\theta}{\displaystyle\int_0^\pi \sin\theta\,d\theta} = 1 + \frac{\alpha}{1-\alpha}\ln\alpha$$

The moderator in NR is D_2O and assuming naively that the expression we derived holds for a molecule we have $A = 20$ and $\xi = 0.10$. However, an appropriate average of the deuterium and oxygen weighted with the scattering cross-section yields $\xi = 0.51$.

The fission neutrons have an average energy of 1 MeV and the average temperature of the moderator is 353 K = 0.03042 eV. So the number of collisions required to reduce the average energy of the fission neutron from 1 MeV to 0.03042 eV is:

$$n = \frac{\ln\dfrac{E_0}{E_n}}{\xi} = \frac{\ln\left(\dfrac{10^6\,\text{eV}}{0.03042\,\text{eV}}\right)}{0.51} \simeq 34$$

We are now in a position to estimate the distance between two fuel channels. For simplicity, consider a neutron beam traveling along the x-axis and undergoing only forward scattering. These assumptions will give only a very crude estimate of the distance between two fuel channels. The intensity of the beam I(x) is attenuated, $I(x) = I(0)e^{-\Sigma_s x}$, where I(0) is the intensity at x = 0 and Σ_s represents the scattering constant. This attenuation formula holds for light, and indeed for any beam phenomena where scatterings/absorptions are independent events. Let x be the distance traveled by a neutron between two scatterings. Then $\bar{x}$, the average distance a neutron travels without any scattering is:

$$\bar{x} = \frac{1}{I(0)} \int_0^\infty I(x)\,dx = \frac{1}{\Sigma_s}$$

So the average distance a neutron travels in the moderator after it leaves a fuel channel is $n\bar{x} = n/\Sigma_s$. Since the neutron travels in three dimensions, the distance traveled in one dimension is approximately $n/(3\Sigma_s)$, which is also an estimate of the distance between two fuel channels. Substituting n = 34 and $\Sigma_s = 45$ m^{-1} for D$_2$O yields the distance between two fuel channels to be 0.262 m.

Nuclear Reactor

In the steady state, the NR maybe operated at any constant neutron flux ψ provided the leakage of neutrons from it is compensated by an excess production of neutrons in it. For a cylindrical reactor, an elaborate calculation yields that the leakage rate is:

$$k_1 \left[\left(\frac{2.405}{R} \right)^2 + \left(\frac{\pi}{H} \right)^2 \right] \psi$$

The excess production rate is $k_2 \psi$ which is proportional to the flux. The constants k_1 and k_2 depend on the material properties of the NR. Let us consider an NR with $k_1 = 1.021 \times 10^{-2}$ m and $k_2 = 8.787 \times 10^{-3}$ m^{-1}. Noting that for a fixed volume the leakage rate is to be minimized for efficient fuel utilization, we can obtain the dimensions of the NR in the steady state. For constant volume V = $\pi R^2 H$,

$$\frac{d}{dH} \left[\left(\frac{2.405}{R} \right)^2 + \left(\frac{\pi}{H} \right)^2 \right] = \frac{d}{dH} \left[\frac{2.405^2 \pi H}{V} + \frac{\pi^2}{H^2} \right]$$

$$= \frac{2.405^2 \pi}{V} - 2\frac{\pi^2}{H^3} = 0,$$

gives $\left(\frac{2.405}{R} \right)^2 = 2\left(\frac{\pi}{H} \right)^2$.

For steady state,

$$1.021 \times 10^{-2} \left[\left(\frac{2.405}{R} \right)^2 = \left(\frac{\pi}{H} \right)^2 \right] \psi = 8.787 \times 10^{-3}\, \psi$$

This yields H = 5.866 m and R = 3.175 m. The height and radius are 5.940 m and 3.192 m respectively.

Finally, we will estimate the critical mass M_F of UO_2 required to operate the NR in steady state. For this, we need to find the number of fuel channels F_n present in the critical volume of the NR calculated above. Though this can be done by rigour, we adopt a quick and easier method. The fuel channels are in a square arrangement with nearest neighbour distance 0.286 m; so the effective area per channel is 0.286^2 m² = 8.180×10^{-2} m². The cross-sectional area of the core is πR^2 = 3.142 $\times (3.175)^2$ = 31.67 m²; so the maximum number of fuel channels that can be accommodated in the cylinder is the integer part of 31.67/0.0818 = 387. The effective radius of a fuel channel (if it were solid) is 3.617×10^{-2} m. So the mass of fuel,

$$M_F = 387 \times \left(\pi \times 0.03617^2 \times 5.866 \right) \times 10600 = 9.892 \times 10^4 \, \text{kg}$$

The total volume of the fuel is 387 × (π × 0.03617^2 × 5.866) = 9.332 m³. If the reactor works at 12.5 % efficiency, then using Q (the heat produced per unit volume), we have that the power output of the reactor is 9.332 × 4.917 × 10^8 × 0.125 = 573 MW.

References

- Reactors: whatisnuclear.com, Retrieved 15 January, 2019
- Nuclear-101-how-does-nuclear-reactor-work: energy.gov, Retrieved 16 April, 2019
- Nuclear-reactors: large.stanford.edu, Retrieved 18 March, 2019
- Atoms Forge a Scientific Revolution, Argonne National Laboratory, 10 July 2012.

5

Nuclear Fuel: Manufacturing and Reprocessing

Nuclear fuel is one of the important components of a nuclear reactor. Some of its aspects include fabrication of nuclear fuel, energy density calculations of nuclear fuel, nuclear fuel cycle, reprocessing of nuclear fuel, etc. This chapter closely examines these aspects of nuclear fuel manufacturing and reprocessing, to provide an extensive understanding of the subject.

Nuclear Fuel

Nuclear fuel is the fuel that is used in a nuclear reactor to sustain a nuclear chain reaction. These fuels are fissile, and the most common nuclear fuels are the radioactive metals uranium-235 and plutonium-239. All processes involved in obtaining, refining, and using this fuel make up a cycle known as the nuclear fuel cycle.

Uranium-235 is used as a fuel in different concentrations. Some reactors, such as the CANDU reactor, can use natural uranium with uranium-235 concentrations of only 0.7%, while other reactors require the uranium to be slightly enriched to levels of 3% to 5%. Plutonium-239 is produced and used in reactors (specifically fast breeder reactors) that contain significant amounts of uranium-238. It can also be recycled and used as a fuel in thermal reactors. Current research is being done to investigate how thorium-232 can be used as a fuel.

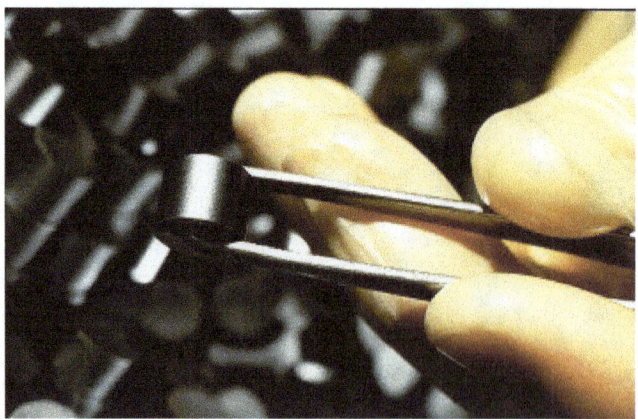

Production

Fuel fabrication plants are facilities that convert enriched uranium into fuel for nuclear reactors. For light water reactors, uranium is received from an enrichment plant in solid form. It is then converted into a gas and chemically converted into a uranium dioxide powder. This powder is then pressed into pellets and packed into fuel assemblies. A mixed oxide fuel can also be created when the uranium powder is packed along with plutonium oxide. The hazards present at fuel fabrication facilities—mainly chemical and radiological—are similar to the hazards at enrichment plants. These facilities generally pose a low risk to the public.

Use

When used in a reactor, the fuels used can have a variety of different forms a metal, an alloy, or some sort of oxide. Most nuclear reactors are fueled with a compound known as uranium dioxide. This uranium dioxide is put together in a fuel assembly and inserted into the nuclear reactor—where it can stay for several months or up to a few years. While in the reactor the fuel undergoes nuclear fission and releases energy. This released energy is used to generate electricity. Neutrons released during the fission process allow for a fission chain reaction to occur, allowing energy to be generated continually. The fuel is removed from the reactor after large amounts of the fuel—whether it is uranium-235 or plutonium-239—have undergone fission. The "used" nuclear fuel is known as spent or irradiated fuel. After use, the fuel must be cooled for a few years as it is extremely hot.

The spent fuel is placed in large, deep pools of water that act as a coolant and a radiation shield. The coolant property allows the water to remove the decay heat and the shielding abilities protect workers from the radioactivity of the fuel. After cooling, the fuel can be re-purposed or sent to storage depending on regulations.

Fuel Assembly

Nuclear reactors are powered by powdered uranium dioxide that has been compressed into small pellets, shown in figure. However, a power plant requires *many* of these pellets to run. Thus large numbers of these pellets are bundled into a fuel rod. A single uranium fuel pellet, only as big as a fingertip, contains as much energy as 481 cubic meters of natural gas, 807 kilograms of coal or 564 liters of oil. These rods are composed of numerous pellets of fissionable uranium fuel and can be several meters in length and about a centimeter in diameter. Then several of these rods, generally a dozen or more, are held together by strong metallic brackets in a fuel assembly. These rods are not bunched tightly together, rather there are several millimeters between each rod to allow coolant to flow between them. The tubes containing the pellets of uranium are generally composed of zirconium.

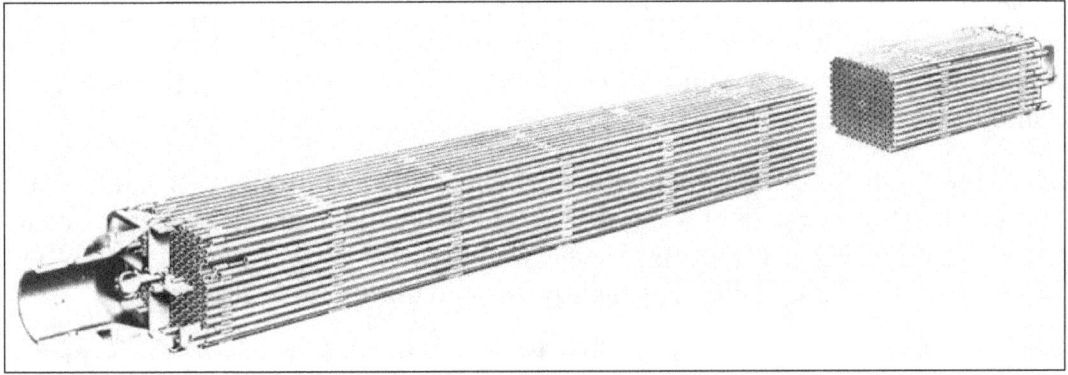

A nuclear fuel bundle.

Advantages and Disadvantages of Nuclear Fuels

Worldwide, there are extensive reserves of uranium left to mine. While nuclear fuel is not renewable, it is sustainable since there is so much of it. It will run out eventually, but not for centuries. Unlike fossil fuels, using nuclear fuels to produce energy does not directly produce carbon dioxide or sulfur dioxide. It should be mentioned that the processes of mining, transporting, and refining the fuel have carbon emissions associated with them, comparable to those of wind and solar power. Although the carbon footprint of using nuclear fuels is smaller, there are still disadvantages of using nuclear fuel. The waste, while a much lower volume must be handled very carefully because of its radioactivity. Nuclear fuels require far more complicated systems to extract their energy, which calls for greater regulation. These complex systems and regulation make for very long build times. In addition, public opinions on nuclear energy tend to be more negative than with other energy sources. The over-estimation of the dangers associated with releases of radioactive material is a significant issue, as large-scale nuclear incidents are rare.

Nuclear Fuel and its Fabrication

- Fuel fabrication is the last step in the process of turning uranium into nuclear fuel rods.

- Batched into assemblies, the fuel rods form the majority of a reactor core's structure.

- This transformation from a fungible material – uranium – to high-tech reactor components is conceptually different from the refining and preparation of fossil fuels.

- Nuclear fuel assemblies are specifically designed for particular types of reactors and are made to exacting standards.

- Utilities and fabricators have collaborated to greatly improve fuel assembly performance, and an international accident-tolerant fuel program is under way.

While all present fuel is oxide, R&D is focused on metal, nitride and other forms. The first modern metal fuel is due to be trialled in commercial reactors.

Nuclear reactors are powered by fuel containing fissile material. The fission process releases large amounts of useful energy and for this reason the fissioning components – U-235 and Pu-239 – must be held in a robust physical form capable of enduring high operating temperatures and an intense neutron radiation environment. Fuel structures need to maintain their shape and integrity over a period of several years within the reactor core, thereby preventing the leakage of fission products into the reactor coolant.

The standard fuel form comprises a column of ceramic pellets of uranium oxide, clad and sealed into zirconium alloy tubes. For light water reactor (LWR) fuel, the uranium is enriched to various levels up to about 4.8% U-235. Pressurised heavy water reactor (PHWR) fuel is usually unenriched natural uranium (0.7% U-235), although slightly-enriched uranium is also used.

Fuel assembly performance has improved since the 1970s to allow increased burn-up of fuel from 40 GWday/tU to more than 60 GWd/tU. This is correlated with increased enrichment levels from about 3.25% to 5% and the use of advanced burnable absorber designs for PWR, using gadolinium. Core monitoring giving detailed real-time information has enabled better fuel performance also.

The fabrication of fuel structures – called assemblies or bundles – is the last stage of the front end of the nuclear cycle shown in figure, and represents less than 20% of the final cost of the fuel. The process for uranium-plutonium mixed oxide (MOX) fuel fabrication is essentially the same – notwithstanding some specific features associated with handling the plutonium component.

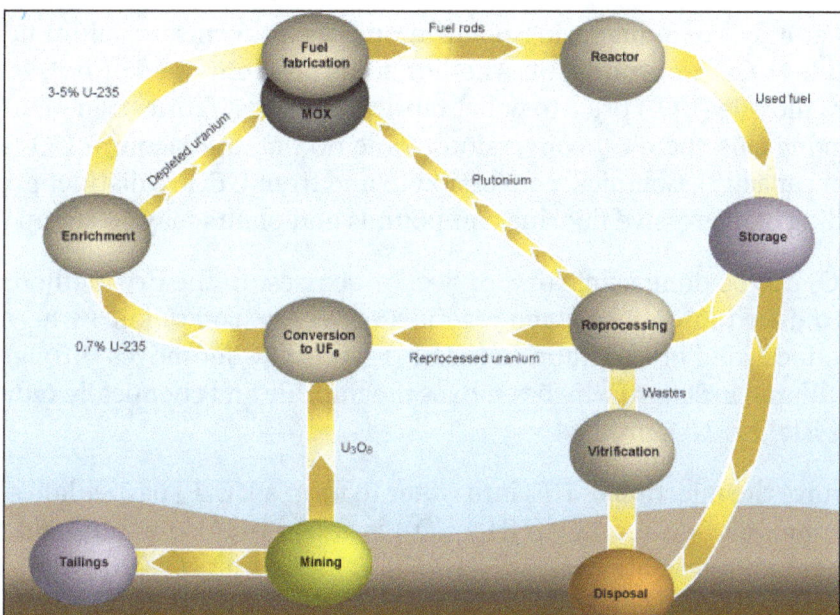

The closed nuclear fuel cycle, showing primary and recycled materials flow.

The industry is dominated by four companies serving international demand for light water reactors: Areva, Global Nuclear Fuel (GNF), TVEL and Westinghouse. GNF is mostly for BWR, and TVEL for PWR.

Process of Nuclear Fuel Fabrication

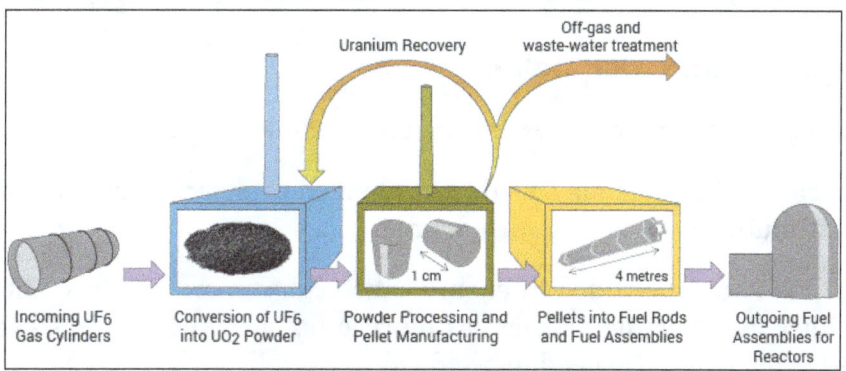

The fuel fabrication process.

There are three main stages in the fabrication of the nuclear fuel structures used in LWRs and PHWRs:

- Producing pure uranium dioxide (UO_2) from incoming UF_6 or UO_3.

- Producing high-density, accurately shaped ceramic UO_2 pellets.

- Producing the rigid metal framework for the fuel assembly – mainly from zirconium alloy; and loading the fuel pellets into the fuel rods, sealing them and assembling the rods into the final fuel assembly structure.

UO_2 Powder Production

Uranium arrives at a fuel manufacturing plant in one of two forms, uranium hexafluoride (UF_6) or uranium trioxide (UO_3), depending on whether it has been enriched or not. It needs to be converted to uranium dioxide (UO_2) prior to pellet fabrication. Most fabrication plants have their own facilities for effecting this chemical conversion (some do not, and acquire UO_2 from plants with excess conversion capacity). Chemical conversion to and from UF_6 are distinct processes, but both involve the handling of aggressive fluorine compounds and plants may be set up to do both.

Conversion to UO_2 can be done using 'dry' or 'wet' processes. In the dry method, UF_6 is heated to a vapour and introduced into a two stage reaction vessel (eg, rotary kiln) where it is first mixed with steam to produce solid uranyl fluoride (UO_2F_2) – this powder moves through the vessel to be reacted with H_2 (diluted in steam) which removes the fluoride and chemically reduces the uranium to a pure microcrystalline UO_2 product.

Wet methods involve the injection of UF_6 into water to form a UO_2F_2 particulate slurry. Either ammonia (NH_3) or ammonium carbonate ($(NH_3)_2CO_3$) is added to this mixture and the UO_2F_2 reacts to produce; ammonium diuranate (ADU, $(NH_3)_2U_2O_7$) in the first case, or ammonium uranyl carbonate (AUC, $UO_2CO_3.(NH_3)_2CO_3$) in the latter case. In both cases the slurry is filtered, dried and heated in a reducing atmosphere to pure UO_2. The morphology of UO_2 powders deriving from the ADU and AUC routes are different, and this has a bearing on final pellet microstructure.

Wet methods are slightly more complex and give rise to more wastes, however, the greater flexibility in terms of UO_2 powder properties is an advantage.

For the conversion of UO_3 to UO_2, water is added to UO_3 so that it forms a hydrate. This solid is fed (wet or dry) into a kiln operating with a reducing atmosphere and UO_2 is produced.

Manufacture of Ceramic UO_2 pellets

The UO_2 powder may need further processing or conditioning before it can be formed into pellets:

- Homogenization: Powders may need to be blended to ensure uniformity in terms of particle size distribution and specific surface area.

- Additives: U_3O_8 may be added to ensure satisfactory microstructure and density for the pellets. Other fuel ingredients, such as lubricants, burnable absorbers (e.g. gadolinium) and pore-formers may also need to be added.

Conditioned UO_2 powder is fed into dies and pressed biaxially into cylindrical pellet form using a load of several hundred MPa – this is done in pressing machines operating at high speed. These 'green' pellets are then sintered by heating in a furnace at about 1750 °C under a precisely controlled reducing atmosphere (usually argon-hydrogen) in order to consolidate them. This also has the effect of decreasing their volume. The pellets are then machined to exact dimensions – the scrap from which being fed back into an earlier stage of the process. Rigorous quality control is applied to ensure pellet integrity and precise dimensions.

For most reactors pellets are just under one centimetre in diameter and a little more than one centimetre long. A single pellet in a typical reactor yields about the same amount of energy as one tonne of steaming coal.

Burnable absorbers (or burnable 'poisons') such as gadolinium may be incorporated (as oxide) into the fuel pellets of some rods to limit reactivity early in the life of the fuel. Burnable absorbers have a very high neutron absorption cross-section and compete strongly for neutrons, after which they progressively 'burn-out' and convert into nuclides with low neutron absorption leaving fissile (U-235) to react with neutrons. Burnable absorbers enable longer fuel life by allowing higher fissile enrichment in fresh fuel, without excessive initial reactivity and heat being generated in the assembly.

Gadolinium, mostly at up to 3g oxide per kilogram of fuel, requires slightly higher fuel enrichment to compensate for it, and also after burn-up of about 17 GWd/t it retains about 4% of its absorbtive effect and does not decrease further. Zirconium diboride integral fuel burnable absorber (IFBA) as a thin coating on normal pellets burns away more steadily and completely, and has no impact on fuel pellet properties. It is now used in most US reactors and a few in Asia. China has this technology for AP1000 reactors.

Manufacture and Loading of the Fuel Assembly Framework

Nuclear fuel designs dictate that the pellet-filled rods have a precise physical arrangement in terms of their lattice pitch (spacing), and their relation to other features such as water (moderator) channels and control-rod channels. The physical structures for holding the fuel rods are therefore engineered with extremely tight tolerances. They must be resistant to chemical corrosion, high temperatures, large static loads, constant vibration, fluid and mechanical impacts. Yet they must also be as neutron-transparent as possible.

Assembly structures comprise a strong framework made from steel and zirconium upon which are fixed numerous grid support pieces that firmly hold rods in their precise lattice positions. These are made from zirconium alloy and must permit (and even enhance) the flow of coolant water around the fuel rod. The grid structures grip the fuel rod and so are carefully designed to minimise the risk of vibration-induced abrasion on the cladding tube – called 'fretting' wear.

All fuel fabricators have highly sophisticated engineering processes and quality control for the timely manufacture of their assembly structures.

Pellets meeting QA specifications are loaded into tubes made from an appropriate zirconium alloy, referred to as the 'cladding'. The filled tube is flushed with helium and pressurized with tens of atmospheres (several MPa) of this gas before the ends are sealed at each end by precision welding. A free space is left between the top of the pellet stack and the welded end-plugs – this is called the 'plenum' space and it accommodates thermal expansion of the pellets and some fission product gases. A spring is usually put into the plenum to apply a compressive force on the pellet stack and prevent its movement.

The completed fuel rods are then fixed into the prefabricated framework structures that hold the rods in a precisely defined grid arrangement.

In order to maximize the efficiency of the fission reaction the cladding and indeed all other structural parts of the assembly must be as transparent as possible to neutrons. Different forms of zirconium alloy, or zircaloy, are therefore the main materials used for cladding. This zircaloy includes small amounts of tin, niobium, iron, chromium and nickel to provide necessary strength and corrosion resistance. Hafnium, which typically occurs naturally with zirconium deposits, needs to be removed because of its high neutron absorption cross-section. The exact composition of the alloy used depends on the manufacturer and is an important determiner in the quality of the fuel assembly. Zircaloy oxidizes in air and water, and therefore it has an oxidized layer which does not impair function.

Rigorous quality control measures are employed at all process points in order to ensure traceability of all components in case of failures.

The major process safety concerns at nuclear fuel fabrication facilities are those of fluoride handling and the risk of a criticality event if insufficient care is taken with the arrangement of fissile materials. Both risks are managed through the rigorous control of materials, indeed, fuel fabrication facilities operate with a strict limitation on the enrichment level of uranium that is handled in the plant – this cannot be higher than 5% U-235, essentially eliminating the possibility of inadvertent criticality.

Types of Nuclear Fuel Assemblies for Different Reactors

There is considerable variation among fuel assemblies designed for the different types of reactor. This means that utilities have limited choice in suppliers of fabricated fuel assemblies, especially for PWRs.

PWR Fuel

Pressurised water reactors (PWRs) are the most common type of nuclear reactor accounting for two-thirds of current installed nuclear generating capacity worldwide. A PWR core uses normal water as both moderator and primary coolant – this is kept under considerable pressure (about 10 MPa) to prevent it from boiling, and its temperature rises to about 330°C after its upward passage past the fuel. It then goes through massive pipes to a steam generator.

Fuel for western PWRs is built with a square lattice arrangement and assemblies are characterized by the number of rods they contain, typically, 17×17 in current designs. A PWR fuel assembly stands between four and five metres high, is about 20 cm across and weighs about half a tonne. The assembly has vacant rod positions – space left for the vertical insertion of a control rod. Not every assembly position requires fuel or a control rod, and a space may be designated as a "guide thimble" into which a neutron source rod, specific instrumentation, or a test fuel segment can be placed.

A PWR fuel assembly comprises a bottom nozzle into which rods are fixed through the lattice and to finish the whole assembly it is crowned by a top nozzle. The bottom and top nozzles are heavily constructed as they provide much of the mechanical support for the fuel assembly structure. In the finished assembly most rod components will be fuel rods, but some will be guide thimbles, and one or more are likely to be dedicated to instrumentation. A PWR fuel assembly is shown in figure. PWR fuel assemblies are rather uniform compared with BWR ones and those in any particular reactor must have substantially the same design.

An 1100 MWe PWR core may contain 193 fuel assemblies composed of over 50,000 fuel rods and some 18 million fuel pellets. Once loaded, fuel stays in the core for several years depending on the design of the operating cycle. During refuelling, every 12 to 18 months, some of the fuel - usually one third or one quarter of the core – is removed to storage, while the remainder is rearranged to a location in the core better suited to its remaining level of enrichment.

Russian PWR reactors are usually known by the Russian acronym VVER. Fuel assemblies for these are characterized by their hexagonal arrangement, but are otherwise of similar length and structure to other PWR fuel assemblies. Most is made by TVEL in Russia, but Westinghouse in Sweden also fabricates it and is increasing capacity to do so. TVEL is instigating using erbium as a burnable poison in fuel enriched to about 6.5% in order to prolong the intervals between refuelling to two years.

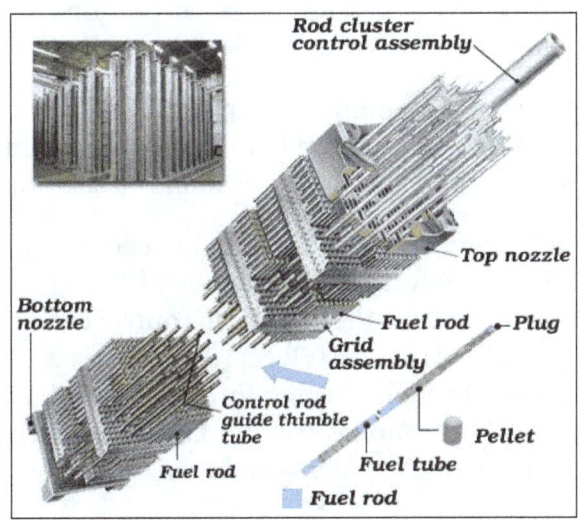

Schematic view of PWR fuel assembly.

A PWR fuel assembly.

A VVER-1000 fuel assembly.

BWR Fuel

Boiling water reactors (BWRs) are the second most common nuclear reactor type accounting for almost one-quarter of installed nuclear generating capacity. In a boiling water reactor water is turned directly to steam in the reactor pressure vessel at the top of the core and this steam (at about 290 °C and 7 MPa) is then used to drive a turbine.

BWRs also use fuel rods comprising zirconium-clad uranium oxide ceramic pellets. Their arrangement into assemblies is again based on a square lattice, with pin geometries ranging from 6x6 to 10x10 or 11×11. Fuel life and management strategy is similar to that for a PWR.

But BWR fuel is fundamentally different from PWR fuel in certain ways: (i) Four fuel assemblies and a cruciform shaped control blade form a 'fuel module', (ii) each assembly is isolated from its neighbours by a water-filled zone in which the cruciform control rod blades travel (they are inserted from the bottom of the reactor), (iii) each BWR fuel assembly is enclosed in a zircaloy sheath or channel box which directs the flow of coolant water through the assembly and during this passage it reaches boiling point, (iv) BWR assemblies contain larger diameter water channels – flexibly designed to provide appropriate neutron moderation in the assembly.

The zircaloy tubes are allowed to fill with water thus increasing the amount of moderator in the central region of the assembly. Different enrichment levels are used in the rods in varying positions – lower enrichments in the outer rods, and higher enrichments near the centre of the bundle. A BWR reactor is designed to operate with 12-15% of the water in the top part of the core as steam, and hence with less moderating effect and thus efficiency there.

For many BWR models, control of reactivity to enable load-following can be achieved by changing the rate of circulation inside the core. Jet pumps located in the annulus between the outer wall of the vessel and an inner wall called the shroud increase the flow of water up through the fuel assembly. At high flow rates steam bubbles are removed more quickly, and hence moderation and reactivity is increased. When flow rate is decreased, moderation decreases as steam bubbles are present for longer and hence reactivity drops. This allows for a variation of about 25% from the maximum rated power output, enabling load-following more readily than with a PWR.

Control rods are used when power levels are reduced below 75%, but they are not part of the fuel assembly as in a PWR. They are bottom-entry – pushed upwards so that rods intercept the lower, more reactive, zone of the fuel assemblies first.

BWR fuel fabrication takes place in much the same way as PWR fuel.

A cross sectional diagram of a BWR assembly is shown in figure. BWR fuel assemblies therefore operate more as individual units, and different designs may be mixed in any core load, giving more flexibility to the utility in fuel purchases.

GE's Global Nuclear Fuels is developing fuel with new clad material – NSF – containing 1% niobium, 1% tin, 0.35% iron (Nb, Sn, Fe) to reduce or eliminate fuel channel distortion due to chemical interaction with zircaloy and in 2013, 8% of cores were using this. Toshiba and ceramics company Ibiden in Japan are developing silicon carbide sheaths or channel boxes for BWR fuel assemblies.

Westinghouse plans to produce lead test assemblies of its TRITON11 fuel (11×11 configuration) for BWRs in 2019. This has a low-tin zirconium channel material and new fuel cladding. It says that this fuel features improved economy, robust mechanical design and high-performing material. It has been optimised for both short- and long-cycle operation as well as for uprated cores and higher burn-ups.

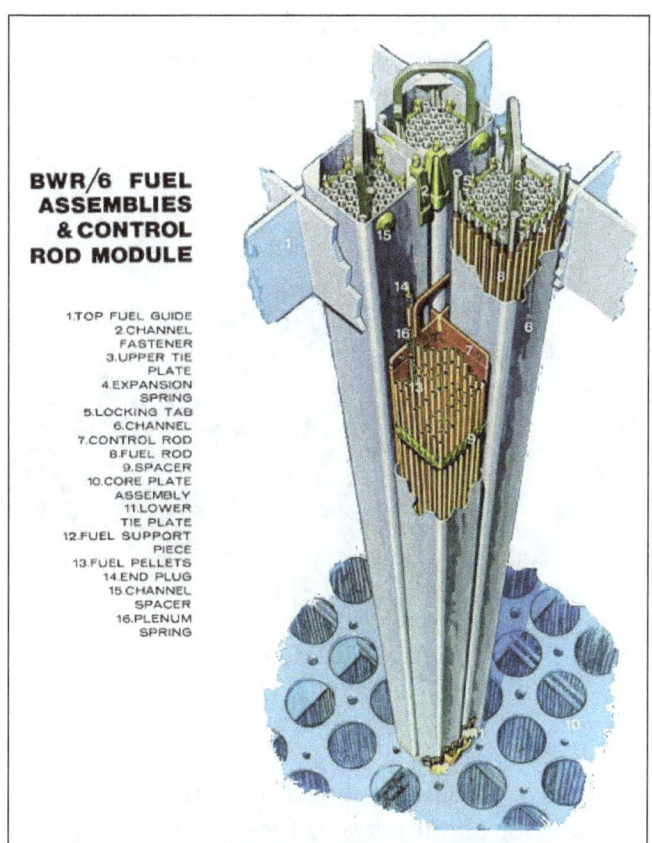

Schematic view of BWR fuel assembly.

Pressurised Heavy Water Reactors (CANDU) Fuel

Pressurised Heavy Water Reactors (PHWRs) are originally a Canadian design (also called "CANDU") accounting for ~6% of world installed nuclear generating capacity. PHWRs use pressure tubes in which heavy water moderates and cools the fuel. They run on natural (unenriched) or slightly-enriched uranium oxide fuel in ceramic pellet form, clad with zirconium alloy.

PHWR fuel rods are about 50 cm long and are assembled into 'bundles' approximately 10 cm in diameter. A fuel bundle comprises 28, 37 or 43 fuel elements arranged in several rings around a central axis. Their short length means that they do not require the support structures characteristic of other reactor fuel types. PHWR fuel does not attain high burn-up, nor does it reside in the reactor core for very long and so the fuel pellets swell very little during their life. This means that PHWR fuel rods do not need to maintain a pellet-cladding gap, nor be highly pressurized with a filling gas (as for LWR fuel), indeed, the metal cladding is allowed to collapse onto the fuel pellet thereby assuring good thermal contact.

The fuel bundles are loaded into horizontal channels or pressure tubes which penetrate the length of the reactor vessel (known as the calandria), and this can be done while the reactor is operating at full power. About twelve bundles are loaded into each fuel channel depending on the model – a 790 MWe CANDU reactor contains 480 fuel channels composed of 5,760 fuel bundles containing over 5 million fuel pellets.

The on-load refueling is a fully-automated process: new fuel is inserted into a channel at one end and used fuel is collected at the other. This feature means that the PHWR is inherently flexible with its fuel requirements, and can run on different fuels requiring different residence times, eg natural uranium, slightly enriched uranium, plutonium-bearing fuels and thorium-based fuels.

PHWR fuel bundles.

AGR Fuel

The Advanced Gas-Cooled Reactor (AGR) is a second-generation UK-designed nuclear reactor only used in UK. AGRs account for about 2.7% of total global nuclear generating capacity. They employ a vertical fuel channel design, and use CO_2 gas – a very weak moderator – as the primary coolant.

AGR fuel assemblies consist of a circular array of 36 stainless steel clad fuel pins each containing 20 enriched UO_2 fuel pellets, and the assembly weighs about 43 kilograms. Enrichment levels vary up to about 3.5%. Stainless steel allows for higher operating temperatures but sacrifices some neutron economy. The assembly is covered with a graphite sheath which serves as a moderator. Eight assemblies are stacked end on end in a fuel channel, inserted down through the top of the reactor. During refueling this whole stack is replaced. Fuel life is about five years, and refueling can be carried out on-load through a refuelling machine.

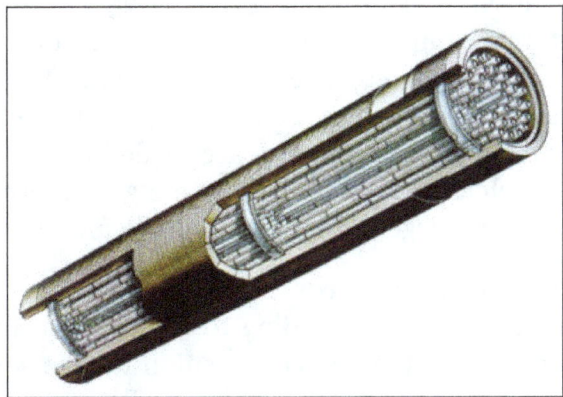

cutaway of an AGR fuel assembly.

RBMK Fuel

The RBMK reactor is an early Soviet design, developed from plutonium production reactors. Eleven units are in operation (3% of world total), with control systems and oxide fuel greatly modified since 1990. It employs vertical pressure tubes (just under1700 of these, each about 7 metres long) running through a large graphite moderator. The fuel is cooled by light water water, which is allowed to boil in the primary circuit, much as in a BWR.

RBMK fuel rods are about 3.65 metres long, and a set of 18 forms a fuel bundle about 8 cm diameter. Two bundles are joined together and capped at either end by a top and bottom nozzle, to form a fuel assembly with an overall length of about 10 metres, weighing 185 kilograms. Since 1990 RBMK fuel has had a higher enrichment level, increasing from about 2% to average 2.8% (varying along the fuel element from 2.5% to 3.2%) and it now includes about 0.6% erbium (a burnable absorber). This has the effect of improving overall safety and increasing fuel burn-up. This new fuel can stay in the reactor for periods of up to six years before needing to be removed. All RBMK reactors now use recycled uranium from VVER reactors.

As with other pressure tube designs such as the PHWR, the RBMK reactor is capable of on-load refueling.

Fast Neutron Reactor Fuel

There is only one commercially operating fast reactor (FNRs) in service today – the BN-600 at Beloyarsk in Russia. There are two FNRs under construction – a 800 MWe unit in Russia and a 500 MWe unit in India (which expects to build five more). Two BN-800 units were planned in China.

Fast neutron reactors (FNR) are unmoderated and use fast neutrons to cause fission. Hence they mostly use plutonium as their basic fuel, or sometimes high-enriched uranium to start them off (they need about 20-30% fissile nuclei in the fuel). The plutonium is bred from U-238 during operation. If the FNR is configured to have a conversion ratio above 1 (ie more fissile nuclei are created than fissioned) as originally designed, it is called a Fast Breeder Reactor (FBR). FNRs use liquid metal coolants such as sodium and operate at higher temperatures.

Apart from the main FNR fuel, there are numerous heavy nuclides - notably U-238, but also Am, Np and Cm that are fissionable in the fast neutron spectrum – compared with the small number of fissile nuclides in a slow (thermal) neutron field (just U-235, Pu-239 and U-233). A FNR fuel can therefore include a mixture of transuranic elements. Also it can be in one of several chemical forms, including; standard oxide ceramic, mixed oxide ceramic (MOX), single or mixed nitride ceramics, carbides and metallic fuels. Further, FNR fuel can be fabricated in pellet form or using the 'vibro-pack' method in which graded powders are loaded and compressed directly into the cladding tube. Carbide fuels such as used in India have a higher thermal conductivity than oxide fuels and can attain breeding ratios larger than those of oxide fuels but less than metal fuels.

The core of an FNR is much smaller than a conventional reactor, and cores tend to be designed with distinct 'seed' and 'blanket' regions according to whether the reactor is to be operated as a 'burner' or a 'breeder'. In each case the fuel composition for the seed and blanket regions are different – the central seed region uses fuel with a high fissile content (and thus high power and neutron emission level), and the blanket region has a low fissile content but a high level of neutron absorbing material which can be fertile (for a breeding design, e.g. U-238) or a waste absorber to be transmuted.

BN-600 fuel assemblies are 3.5 m long, 96 mm wide, weigh 103 kg and comprise top and bottom nozzles (to guide coolant flow) and a central fuel bundle. The central bundle is a hexagonal tube and for seed fuel houses 127 rods, each 2.4 m long and 7 mm diameter with ceramic pellets in three uranium enrichment levels; 17%, 21% and 26%. Blanket fuel bundles have 37 rods containing depleted uranium. BN-600 fuel rods use low-swelling stainless steel cladding.

FNRs use liquid metal coolants such as sodium or a lead-bismuth eutectic mixture and these allow for higher operating temperatures – about 550 °C, and thus have higher energy conversion efficiency. They are capable of high fuel burn-up.

Nuclear Fuel Performance in Reactors

Nuclear fuel operates in a harsh environment in which high temperature, chemical corrosion, radiation damage and physical stresses may attack the integrity of a fuel assembly. The life of a fuel assembly in the reactor core is therefore regulated to a burn-up level at which the risk of its failure is still low. Fuel 'failure' refers to a situation when the cladding has been breached and radioactive material leaks from the fuel ceramic (pellet) into the reactor coolant water. The radioactive materials with most tendency to leak through a cladding breach into the reactor coolant are fission-product gases and volatile elements, notably; krypton, xenon, iodine and caesium.

Fuel leaks do not present a significant risk to plant safety, though they have a big impact on reactor operations and (potentially) on plant economics. For this reason, primary coolant water is monitored continuously for these species so that any leak is quickly detected. The permissible level of released radioactivity is strictly regulated against specifications which take into account the continuing safe operation of the fuel. Depending on its severity a leak will require different levels of operator intervention:

- Very minor leak: No change to operations – the faulty fuel assembly with leaking rod(s) is removed at next refuelling, inspected, and possibly re-loaded.

- Small leak: Allowable thermal transients for the reactor are restricted. This might prevent the reactors from being able to operate in a 'load-follow' mode and require careful monitoring of reactor physics. The faulty fuel assembly with leaking rod is generally removed and evaluated at the next scheduled refuelling.

- Significant leak: The reactor is shut down and the faulty assembly located and removed.

A leaking fuel rod can sometimes be repaired but it is more usual that a replacement assembly is needed (this having a matching level of remaining enrichment). Replacement fuel is one cost component associated with failed fuel. There is also the cost penalty and replacement power from having to operate at reduced power or having an unscheduled shutdown. There may also be higher operation and maintenance costs associated with mitigating increased radiation levels in the plant.

Fuel management is a balance between the economic imperative to burn fuel for longer and the need to keep well within failure-risk limits. Improving fuel reliability extends these limits, and therefore is a critical factor in providing margin to improve fuel burn-up.

The nuclear industry has made significant performance improvements reducing fuel failure rates by about 60% in the 20 years to 2006 to an average of some 14 leaks per million rods The reliability drive continues. Industry-wide programs led by the Electric Power Research Institute (EPRI) and the US Institute of Nuclear Power Operations (INPO) have produced guidelines to help eliminate fuel failures. These programs led to the accident-tolerant fuel program.

Fuel failures in US power reactors are rare. As of early 2014, 97% of US nuclear power plants were free of fuel failures, compared with 71% in 2007, according to EPRI. The annual US failure rate is about one in one million (*i.e.* five rods per year). Fuel engineering continues to improve, *e.g.* with more sophisticated debris filters in assembly structures. Utilities themselves impose more rigorous practices to exclude foreign material entering primary coolant water. Global Nuclear Fuel (GNF) in 2013 had two million fuel rods in operation and claimed to have no leakers among them. (In the early 1970s hydriding and pellet-clad interaction caused a lot of leaks. The 1980s saw an order of magnitude improvement).

At the same time there has been a gradual global trend toward higher fuel burnup*. However, higher burnup generally requires higher enrichment levels, and there is a limit on this given the strict criticality safety limitation imposed on fuel fabrication facilities – the maximum uranium enrichment level that can be handled is normally 5% U-235.

Higher burnup does not necessarily mean better energy economics. Utilities must carefully balance the benefits of greater cycle length against higher front-end fuel costs (uranium, enrichment). Refuelling outage costs may also be higher, depending on length, frequency and the core re-load fraction.

An equally important trend in nuclear fuel engineering is to be able to increase the power rating for fuels, ie, how much energy can be extracted per length of fuel rod. Currently this is limited by the material properties of the zirconium cladding.

Fabrication Supply and Demand

The current annual demand for LWR fuel fabrication services is expressed as a requirement for about 7000 tonnes of enriched uranium being made into assemblies, and this is expected to increase to about 9500 t by 2020. Requirements for PHWRs account for an additional 3,000 t/year and the Gas-cooled Reactor market for around 400 t/year.

Requirements for fuel fabrication services will grow roughly in line with the growth in nuclear generating capacity. However, fabrication requirements are also affected by changes in utilities' reactor operating and fuel management strategies, which are partly driven by technical improvements in fuel fabrication itself. For example, LWR discharge burn-ups have increased steadily as improvements in fuel design have made this possible, and this has tended to reduce fabrication demand, as fuel remains in the reactor for a longer period (though there is a limit to how far burn-ups can be pushed without tackling the 5% enrichment limit in place for criticality safety margins at fuel fabrication plants). A parallel industry-wide focus on increasing fuel performance and reliability has also decreased the demand for fuel to replace defective assemblies.

Plans to build many new reactors affect the demand on fabrication capacity in two ways. The demand for reloads increases in line with the new installed reactor capacity, typically 16 to 20 tonnes per year per GW. Additionally the first cores create a temporary peak demand, since the amount required is about three to four times that of a reload batch in currently operated LWRs (and some of the enrichment is less). An average first core enrichment is about 2.8%.

Provision of Fuel Fabrication Worldwide

Fuel fabrication services are not procured in the same way as, for example, uranium enrichment is bought. Nuclear fuel assemblies are highly engineered products, made to each customer's individual specifications. These are determined by the physical characteristics of the reactor, by the reactor operating and fuel cycle management strategy of the utility as well as national, or even regional, licensing requirements.

Most of the main fuel fabricators are also reactor vendors (or owned by them), and they usually supply the initial cores and early reloads for reactors built to their own designs. However the market for LWR fuel has become increasingly competitive and for most fuel types there are now several competing suppliers – most notably perhaps, Russian fabricator TVEL competes to manufacture Western PWR fuel, and Western fuel fabricators can manufacture VVER fuel. Early in 2016, 41% of Ukraine's VVER fuel came from Westinghouse in Sweden. In May 2016 Global Nuclear Fuel – Americas agreed with TVEL to produce its TVS-K fuel design in the USA for Westinghouse PWRs. TVEL also plans to market the fuel in Europe, and has been qualifying lead assemblies at Ringtails in Sweden.

Currently, fuel fabrication capacity for all types of LWR fuel throughout the world considerably exceeds the demand. It is evident that fuel fabrication will not become a bottleneck in the foreseeable supply chain for any nuclear renaissance. The overcapacity is increased by countries such as China, India and South Korea aiming to achieve self-sufficiency.

In May 2014 a European Commission staff report suggested that as a condition of investment, any non-EU reactor design built in the EU should have more than one source of fuel. The EC's May 2014 European Energy Security Strategy urged: "Ideally, diversification of fuel assembly manufacturing should also take place, but this would require some technological efforts because of the different reactor designs." In June 2015 the Euratom Research and Training Programme provided €2 million to Westinghouse and eight European partners "to establish the security of supply of nuclear fuel for Russian-designed reactors in the EU," especially VVER-440 types. Conceptual design was completed in May 2017, based on fuel provided by Westinghouse to Loviisa in 2001-07.

LWR fuel fabrication capacity worldwide is shown in table. The back-conversion capacities are particularly unevenly distributed. For some fabricators it represents a bottleneck. Some fabricators do not have conversion facilities at all and have to buy such services in the market, while others with excess capacity are even sellers of UO2 powder.

Table: World LWR fuel fabrication capacity, tonnes/yr.

	Fabricator	Location	Conversion	Pelletizing	Rod/assembly
Brazil	INB	Resende	160	160	240
China	CNNC	Yibin	400	400	800
		Baotou	600	600	600
France	AREVA NP-FBFC	Romans	1800	1400	1400
Germany	AREVA NP-ANF	Lingen	800	650	650
India	DAE Nuclear Fuel Complex	Hyderabad	48	48	48
Japan	NFI (PWR)	Kumatori	0	360	284
	NFI (BWR)	Tokai-Mura	0	250	250
	Mitsubishi Nuclear Fuel	Tokai-Mura	450	440	440
	Global Nuclear Fuel – Japan	Kurihama	0	750	750
Kazakhstan	Ulba	Ust Kamenogorsk	2000	2000	0
Korea	KNFC	Daejeon	700	700	700
Russia	TVEL-MSZ*	Elektrostal	1500	1500	1560
	TVEL-NCCP	Novosibirsk	450	1200	1200
Spain	ENUSA	Juzbado	0	500	500
Sweden	Westinghouse AB	Västeras	600	600	600
UK	Westinghouse**	Springfields	950	600	860
USA	AREVA Inc	Richland	1200	1200	1200
	Global Nuclear Fuel – Americas	Wilmington	1200	1000	1000
	Westinghouse	Columbia	1500	1500	1500
Total			14,358	15,818	14,582

Table: World PHWR fuel fabrication capacity, tonnes/yr.

	Fabricator	Location	Rod/Assembly
Argentina	CONUAR	Cordoba & Eizeiza	160
Canada	Cameco	Port Hope	1200
	GNF-Canada	Peterborough	1500
China	CNNC China Northern	Baotou	200
India	DAE Nuclear Fuel Complex	Hyderabad	850
Pakistan	PAEC	Chashma	20
Korea	KEPCO	Taejon	400
Romania	SNN	Pitesti	240
Total			4,570

The LWR fuel fabrication industry has rationalized in recent years, including:

- When Westinghouse Electric was bought by Toshiba, Kazatomprom acquired a 10% share of this.

- Global Nuclear Fuels was formed as a joint venture between General Electric, Toshiba and Hitachi, though Toshiba sold its 14% stake to Hitachi in 2018, raising its share to 40%. There are two 'branches' GNF-A (Americas) and GNF-J (Japan) with different ownership structures. It is best-known for BWR fuel.

- Toshiba purchased 52% of Nuclear Fuel Industries (NFI) in Japan, then agreed to buy the balance from Sumitomo (24%) and Furukawa (24%) to make it wholly-owned.

- Mitsubishi Heavy Industries and AREVA (30%) bought into Mitsubishi Nuclear Fuel and created a US fuel fabrication joint venture.

- Kazatomprom and AREVA agreed to build a 1200 t/yr fuel fabrication plant in Kazakhstan.

Secondary Supply from Recycle

Currently about 100 t/yr year of reprocessed uranium (RepU) is produced at MSZ in Elektrostal, Russia (capacity 250 t/yr) for AREVA contracts. One production line in AREVA's plant in Romans, France is licensed to fabricate 150 t of RepU into fuel per year and PWR assemblies of this type have already been delivered to French, Belgian and UK reactors, and an amount of RepU powder has been sent from Russia to Japan. Limited RepU and enriched RepU (ERU) capacity exists elsewhere also.

At present, nearly all commercial MOX fuel is fabricated in AREVA's MELOX plant in Marcoule. With a capacity of 195 tonnes/yr and a good production rate this plant helps not only to save uranium and enrichment demand, but also frees up LWR fabrication capacity in the market.

The UK's Sellafield MOX plant had a designed capacity of 120 t/yr but was downgraded to 40 t/yr and never reached that level of reliable output before being closed down in 2011. Russia's MOX plant at Zheleznogorsk for fast reactors commenced operation in 2015. Japan's Rokkasho-Mura MOX plant is planned to be operational by 2022, and the US MOX plant in Savannah River was due to produce MOX fuel from weapons plutonium but this project has now been terminated.

The MOX fuel market has weakened somewhat recently with the cessation of its use in Belgium, Germany and Switzerland (moratorium), and the continued loading of MOX fuel in Japan has diminished in the aftermath of the Fukushima accident.

Table: World MOX fuel fabrication capacity, tonnes/yr.

France	Areva NC	Marcoule	195	195
India	DAE Nuclear Fuel Complex	Tarapur	50	50
Japan	JAEA	Tokai-Mura	5	5
	JNFL	Rokkasho-Mura	130	130
Russia	MCC	Zheleznogorsk	60	60
Total			440	440

MOX Fuel

Mixed uranium oxide + plutonium oxide (MOX) fuel has been used in about 30 light-water power reactors in Europe and about ten in Japan. It consists of depleted uranium (about 0.2% U-235), large amounts of which are left over from the enrichment of uranium, and plutonium oxide that derives from the chemical processing of used nuclear fuel (at a reprocessing plant). This plutonium is reactor-grade, comprising about one third non-fissile isotopes.

In a MOX fuel fabrication plant the two components are vigorously blended in a high-energy mill which intimately mixes them such that the powder becomes mainly a single 'solid solution' $(U,Pu)O_2$. MOX fuel with about 7% of rector-grade plutonium is equivalent to a typical enriched uranium fuel. The pressing and sintering process is much the same as for UO_2 fuel pellets, though some plastic shielding is needed to protect workers from spontaneous neutron emissions from the Pu-240 component.

Vibropacked MOX (VMOX) fuel is a Russian variant for MOX fuel production in which blended $(U,Pu)O_2$ and UO_2 powders are directly loaded and packed into cladding tubes where they sinter in-situ under their own operating temperature. This eliminates the need to manufacture pellets to high geometric tolerances, which involves grinding and scrap which are more complex to deal with for Pu-bearing fuels. Russian sources say vibropacked fuel is more readily recycled.

REMIX Fuel

REMIX (from Regenerated Mixture) fuel is produced directly from a non-separated mix of recycled uranium and plutonium from reprocessing used fuel, with a LEU (up to 17% U-235) uranium make-up comprising about 20% of the mix. This gives fuel with about 1% Pu-239 and 4% U-235 which can sustain burn-up of 50 GWd/t over four years. The spent REMIX fuel after four years is about 2% Pu-239 and 1% U-235, and following cooling and reprocessing the non-separated uranium and plutonium is recycled again after LEU addition, which compensates for the even isotopes of both elements.

REMIX-fuel can be repeatedly recycled with 100% core load in current VVER-1000 reactors and, correspondingly reprocessed many times – up to five times, so that with three fuel loads in circulation a reactor could run for 60 years using the same fuel, with LEU recharge. REMIX can serve as a replacement for existing reactor fuel, though it is not yet commercialised. The REMIX cycle can be modified from the above figures according to need.

TRISO High Temperature Reactor Fuel

High temperature reactors (HTR) operate at 750 to 950 °C, and are normally helium-cooled. Fuel for these is in the form of TRISO (tristructural-isotropic) particles less than a millimetre in diameter. Each has a kernel (0.5 mm) of uranium oxycarbide (or uranium dioxide), with the uranium enriched up to 20% U-235, though normally less. This is surrounded by layers of carbon and silicon carbide, giving a containment for fission products which is stable up to very high temperatures. Trials at two US laboratories have confirmed that most fission products remain securely in TRISO particles up to about 1800 °C.

There are two ways in which these particles can be arranged in a HTR: in blocks – hexagonal 'prisms' of graphite; or in billiard ball-sized pebbles of graphite encased in silicon carbide, each with about 15,000 fuel particles and 9g uranium. Either way, the moderator is graphite. HTRs can potentially use thorium-based fuels, such as high-enriched or low-enriched uranium with Th, U-233 with Th, and Pu with Th. Most of the experience with thorium fuels has been in HTRs.

The main HTR fuel fabrication plant is at Baotou in China, the Northern Branch of China Nuclear Fuel Element Co Ltd. From 2015 this makes 300,000 fuel pebbles per year for the HTR-PM under construction at Shidaowan. Previous production has been on a small scale in Germany.

In the USA, BWX Technologies at Lynchburg in Virginia is making high-assay low-enriched (HALEU) TRISO fuel on an engineering scale, funded by the US Department of Energy (DOE), and in October 2019 the company announced an expansion to commercial scale within three years.

X-energy has a TRISO pilot fuel fabrication facility at Oak Ridge National Laboratory (ORNL), in Tennessee. In November 2019 X-energy and GNF agreed to set up commercial HALEU TRI-SO production at GNF's Wilmington plant in North Carolina. This is expected to produce TRISO fuel of "significantly higher quality and at costs that are substantially lower than other potential manufacturers." It would potentially supply the US Department of Defense for micro-reactors and NASA for nuclear thermal propulsion. X-energy is building on the TRISO fuel technology developed under the DOE's Advanced Gas Reactor Fuel Qualification Program through two cooperative agreements with the DOE.

X-energy also has agreements with Centrus Energy in the USA to develop TRISO fabrication technology for uranium carbide fuel, and with NFI at Tokai in Japan, which has 400 kgU/yr HTR fuel capacity. NFI is to supply equipment for X-Energy's TRISO-X plant at ORNL.

X-energy is applying for a loan guarantee from the government for commercialization of a TRISO-based fuel supply chain and is expected to submit a licence application for a commercial plant by mid-2021, though this may now be GNF's prerogative.

Other High-assay Low-enriched Fuel

In connection with a number of small modular reactor (SMR) designs, attention is turning to the need for high-assay low-enriched uranium (HALEU), with enrichment levels between 5% and 20% U-235. In the USA, the Department of Energy (DOE) is proposing to convert metallic HALEU into fuel for research and development purposes at Idaho National Laboratory's Materials and Fuels Complex and the Idaho Nuclear Technology and Engineering Center, to support the development of new reactor technologies with higher efficiencies and longer core lifetimes. HALEU may be metallic or oxide.

HALEU can be produced with existing centrifuge technology, but a number of arrangements would need to be made for this, as well as for deconversion and fuel fabrication. New transport containers would also be required as those for today's enriched UF_6 could not be used due to criticality considerations.

Advanced Nuclear Fuel Technology

Fuel development activities in the nuclear industry have largely focused on improving the reliability of standard zirconium-clad uranium oxide fuels. Increasingly, however, R&D effort is being applied to evolutionary fuel forms that can offer significant improvements in terms of safety, waste management and operating economics, as well as allowing the deployment of new types of reactor.

Accident Tolerant Fuel

Accident tolerant fuel (ATF) is a term used to describe new technologies that enhance the safety and performance of nuclear fuel. ATF may incorporate the use of new materials and designs for cladding and fuel pellets.

Framatome, GE/GNF and Westinghouse are all developing ATF concepts with the help of funding from the US Department of Energy (DOE). Since 2012, the DOE has supported the development of ATF concepts through its Enhanced Accident Tolerant Fuel (EATF) programme. Its objective is to develop new cladding and fuel materials that can better tolerate the loss of active cooling in the core, while maintaining or improving fuel performance and economics during normal operations. A priority of the EATF programme is to minimise the generation of hydrogen.

The joint DOE programme uses the Halden research reactor in Norway to test ATF fuel rods, as well as the Advanced Test Reactor (ATR) and the restarted Transient Reactor Test Facility (TREAT) at the DOE's Idaho National Laboratory (INL). In February 2017 the DOE awarded \$10 million to Framatome (then Areva) over two years for phase 2 of the programme, and similar funding is being provided for GE Hitachi and Westinghouse.

Framatome in phase 2 of the ATF programme from 2017 has been developing a nuclear fuel concept, using a chromium-coated zirconium alloy cladding combined with chromia-doped fuel pellets. The fuel is expected to retain fission gases better and improve pellet-cladding interaction, and the cladding will better resist high-temperature oxidation. In June 2018 the DOE announced testing of Framatome's ATF in the Advanced Test Reactor at Idaho National Laboratory. The first full test assemblies were loaded into Southern Nuclear's Vogtle 2 in March 2019. Exelon plans to load two full PROtect fuel assemblies into Calvert Cliffs 2 in March 2021. Entergy is also due to use them in Arkansas 1. These will also have chromium-coated zircalloy cladding and chromia-doped fuel pellets. Framatome's BWR advanced fuel is Atrium. Framatome is also continuing work on a silicon carbide cladding, and plans to use that cladding on chromia-doped pellets in lead test assemblies in about 2022.

GE Hitachi with GNF is developing a ferritic/martensitic steel alloy cladding (*e.g.* Fe-Cr-Al) known as IronClad for conventional UO_2 fuel. The Fe-Cr-Al cladding has better mechanical strength at high temperatures, retains fission gases better than zirconium alloy and has less potential for hydrogen generation in an accident. GNF's advanced fuel rods were the first developed through the ATF programme to be loaded into a commercial reactor during Hatch 1's spring refuelling outage in March 2018. Exelon's Clinton plant will load them in 2019.

Westinghouse in June 2017 launched its EnCore ATF. It is manufacturing the first EnCore test rods, with lead test assembly insertion of these in September 2019, at Exelon's Byron plant. The initial EnCore fuel comprises high-density uranium silicide fuel pellets inside zirconium cladding with a thin coating of chromium making it more robust chemically. (Uranium silicide – U_3Si_2 – fuels for research reactors are being developed at INL also.) In the second phase, the uranium silicide fuel pellets would be in silicon carbide ceramic matrix composite cladding with a melting point of 2800 °C, and these test assemblies could be loaded into a reactor by 2022. The EnCore ATF fuel development has been supported by awards from the DOE to Westinghouse and a group of partners including General Atomics, several DOE national laboratories, Southern Nuclear Operating Company and Exelon.

Westinghouse said that cost savings will arise because the uranium silicide offers up to 20% higher density of uranium and much higher thermal conductivity which does not degrade with irradiation like UO_2, so fewer fuel assemblies need to be replaced during each refuelling outage. In the second phase of EnCore, the higher temperature tolerance of silicon carbide cladding has potential for revised regulatory requirements, and Westinghouse sees this as a "game changer".

After trials of lead test assemblies, Westinghouse intends to make full reload quantities available from 2027. ATFs present a number of manufacturing challenges and, given the neutronic penalties entailed, enrichment to over 5% could be needed, despite the higher density of uranium in the fuel.

Rosatom's fuel company TVEL plans to offer ATF to its customers by the early 2020s. TVEL is developing ATF for use in Rosatom's VVER reactors and in Western PWRs. Prototype assemblies – one VVER and one for western PWR – are being tested at the MIR research reactor at the Research Institute of Atomic Reactors at Dimitrovgrad. TVEL expects to finalise testing late in 2019, at which point it would decide on the technology to be used for its commercial offering. TVEL stated that its ATF fuel has a heat-resistant coating on the zirconium cladding and fuel pellet composition may include molybdenum or uranium disilicide.

The OECD Nuclear Energy Agency's Expert Group on ATFs for Light Water Reactors reviews cladding and core materials focusing on their fundamental properties and behaviour under normal operation and accident conditions, as described above for the US-led program. Both advanced core materials and components, in particular innovative cladding materials (coated and improved Zrbased alloys, SiC and SiC/SiC composites, advanced steels and refractory metals) and nonfuel components (advanced control rods, BWR channel box) are considered. A sub-group focuses on fuel design to address three categories of innovative fuels: improved UO_2, highdensity fuels and coatedparticle fuels such as the HTR fuel described above.

Metal Fuel

Independently of the US DOE and the international ATF program, Lightbridge is developing an advanced metal fuel concept that may have accident tolerant characteristics.

Metal fuels were used in some earlier reactors such as the UK Magnox design, and also in two US fast reactors, with 5-10% zirconium alloyed. But the higher melting point of uranium oxide has made it the preferred fuel in all reactors for half a century.* At least in the USA, metal fuels have not been made since the 1980s.

UO_2 has a very high melting point – 2865 °C (compared with pure uranium metal – 1132 °C).

However, metal has much better thermal conductivity than ceramic oxide, and recent research has turned back to metal fuel forms. Babcock & Wilcox Nuclear Energy was working with Lightbridge to set up a pilot plant for metal fuel which is 50:50 (by mass) Zr-U alloy, with uranium enriched to almost 20% and having a multi-lobed and helically twisted rod geometry. The increased enrichment compensates for reductions in the initial fissile loading and in the derivative plutonium. Melting point of the alloy is about 1600 °C, and average operating temperature in the fuel is up to 370 °C (rather than about 1250 °C in normal oxide fuel), the thermal conductivity being five times better than oxide fuel. BWXT in the USA has now completed its assessment of feasibility and prepared a fabrication plan for manufacturing fuel samples.

Each Lightbridge fuel rod consists of a central displacer of zirconium surrounded by a four-lobed fuel core with the cladding metallurgically bonded to it. For the hexagonal fuel assemblies for VVERs, the fuel rod is three-lobed. The shape of the rod provides increased surface area for heat transfer and the area between the lobes accommodated swelling during irradiation. The rod has greater structural integrity than current tubes with ceramic pellets inside. The twist of about 180° over about a metre means that the rods are self-spacing while giving good flow characteristics. The fuel operates at a higher power density than oxide fuels and the target burn-up is 21 atomic percent, about three times that of oxide fuels. It is suitable for all LWRs, and is expected to give a power uprate of about 17% in existing PWRs, and up to 30% in new ones designed for the higher power density, with longer fuel cycle. In adition to US and Russian patents, in June 2015 the design was patented in South Korea, where Lightbridge sees a "significant potential market", and in July 2017 this patent protection was extended to cover both the metallic four-lobe design and its manufacture from powder. By November 2017 it had been patented in Japan, China (four patents), South Korea, and Canada.

Lightbridge has also agreed with Canadian Nuclear Laboratories for fabrication of such metal fuel at Chalk River in Canada and testing of it there in the NRU reactor. The agreement was expected to see fabrication and characterization of prototype fuel rods using depleted uranium in early 2016, with irradiation fuel samples using enriched uranium made later the same year. Subject to final approval from the Norwegian Radiation Protection Authority, Lightbridge will test the fuel under prototypical PWR conditions in a pressurized water loop of Norway's 25 MWt Halden research reactor (a boiling heavy water reactor). The initial phase of irradiation testing was to begin in 2017 using short samples to evaluate conductivity, and continue for about three years using 70 cm fuel rods to evaluate cladding and swelling. Tests aim to reach the burnup necessary for insertion of lead test assemblies (LTAs) in a commercial power reactor. The final phase of irradiation testing necessary for batch reloads and full cores operating with a 10% power uprate and a 24-month cycle is expected to take an additional two years and be completed while LTAs have begun operating in the core of a commercial power reactor about 2020.

In April 2015 a group of electric utilities representing half of the US nuclear generating capacity wrote to the Nuclear Regulatory Commission formally expressing interest in Lightbridge's metal fuel, saying that they believed the fuel provided opportunities to improve safety and fuel cycle economics significantly. The utilities' Nuclear Utility Fuel Advisory Board (NUFAB) expects to test the fuel in an operating PWR about 2020.

In March 2016 Lightbridge entered into an exclusive joint development agreement with Areva NP to set up a 50:50 joint venture that would develop, fabricate and commercialize fuel assemblies based on the metallic fuel technology. In November it announced an agreement on key terms for the US-based joint venture, creating "a viable and well-defined commercialization path" covering fuel assemblies for most types of light water reactors, including pressurized water reactors (excluding VVERs), boiling water reactors, small modular reactors and research reactors. In September

2017 a binding agreement was signed with Areva Inc (for New NP) to set up the joint venture in North America. The joint venture between Lightbridge and Framatome* was officially launched in January 2018 and assigned the name Enfission. Commercial sales of the fuel are expected by 2026.

Lightbridge is working with four US nuclear utilities, and late in 2016 a letter of intent was signed with one of them for a lead test fuel assembly demonstration in a US commercial reactor, possibly by 2021.

In the USA the Idaho National Laboratory (INL) has been testing metal fuel fabrication by extrusion for TerraPower's so-called travelling wave reactor (TWR). The fuel in this is designed to be rearranged but not replenished for 40 years.

Thorium-uranium Fuel under Development

Since the early 1990s Russia has had a program to develop a thorium-uranium fuel, based at Moscow's Kurchatov Institute and involving the US company Lightbridge Corporation and US government funding to design fuel for Russian VVER-1000 reactors. Whereas normal fuel uses enriched uranium oxide throughout the fuel assembly, the new design has a demountable centre portion and blanket arrangement, with uranium-zirconium metal fuel rods in the centre and uranium-thorium oxide pellets in conventional fuel rods around it. The Th-232 in the blanket captures neutrons to become U-233, which is fissile and is fissioned in situ. Blanket material remains in the reactor for nine years but the centre portion is burned for only three cycles (as in a normal VVER, 3 or 4.5 years depending on the refuelling interval). No reprocessing of the blanket is envisaged, due both to the difficulty of doing so with thorium fuels and the presence of significant U-232 in the blanket. The two-part fuel assembly has the same geometry as a normal VVER one.

A variant of this design uses plutonium-zirconium metal fuel rods in the central seed assembly, and was earlier known as a plutonium incinerator.

Other Fuel Developments

Other fuel technologies that seem particularly promising, and which could be commercially deployed in the foreseeable future include:

- Ceramic or coated zirconium claddings that prevent the adverse interaction between steam and zirconium at very high temperature.

- High thermal conductivity oxide fuel, such as can be achieved by including additives like beryllium oxide (BeO). Higher conductivity provides higher safety margins and can allow higher operating powers.

- Thoria-based fuels, including mixed thorium-plutonium (Th-MOX) fuel which can achieve a high utilization factor for recycled plutonium.

- Other all-metal fuels and annular LWR fuel, allowing more cooling and therefore safe, high power densities for the fuel and improved economics.

- Pelletized coated-particle fuels, aimed at achieving high safety levels for the fuel that can be left in a light water reactor for very long periods, thereby achieving high burn-up of recycled plutonium and actinide waste components.

Energy Density Calculations of Nuclear Fuel

Energy density is how much work a certain amount of fuel is capable of exerting. In terms of electricity, it is basically a measure of how much fuel you have to burn in a power plant to power the same size city. It's like the miles-per-gallon rating of your power plant.

For any fuel, the characteristics of the power generation system affect exactly how much usable energy is extracted. For instance, if a power plant makes heat to be converted to electricity, the thermal efficiency (ϵ_{th}) determines how much of the heat gets converted to electricity. These values vary from around 33% for coal plants to 35-40% for nuclear plants, to above 60% for combined cycle natural gas plants.

Another factor is how complete the fuel is consumed. For example, a wood fire may burn out before all the energy is extracted from the wood. In nuclear power plants, usually only 5-7% of the energy of the fuel is extracted. Furthermore, the fuel has already gone through an enrichment process so only about 1% of the energy of the mined resource is used. Advanced nuclear power plants called breeder reactors such as the liquid metal fast breeder reactor (LMFBR) or the molten salt breeder reactor (MSBR) can extract much more of the mined energy. The fraction of the energy extracted from the fuel in a reactor is called the burnup.

Since so many factors can change how much electricity comes out of a fuel, it's best to compare the various fuels based on their energy density alone. That is, how much energy could be extracted with 100% thermal efficiency and 100% burnup. This allows us to focus in precisely on the topic of energy density for comparisons.

The easy way to compute energy density of nuclear fuels is to figure out how much fission energy can be released from 1 mole of the fuel. The equation for energy density in MJ/kg is:

$$\text{ED} = \frac{\kappa_{fis}[\text{MeV/fission}]N_A[\text{fissions/mol}]}{A[\text{g/mol}]} \times \frac{1.60217 \times 10^{-19}[\text{Mega Joules/MeV}]}{0.001\text{kg/g}} = \text{MJ/kg}$$

where,

k_{fis} is the energy release per fission for the nuclide of interest. These values are measured by scientists and collected in nuclear data files such as the ones available at the National Nuclear Data Center. Look for the Interpreted field for each nuclide here.

N_A is Avogadro's number, or 6.022e23. This is the number of atoms per mole. Since we're assuming 100% of atoms fission, this is equal to the number of fissions per mole.

A is the atomic mass of the nuclide of interest. This can be found on any Chart of the Nuclides, like this one.

Some Nuclear Energy Density Calculations

Here's how much energy is contained in certain amounts of nuclear fuel.

Table: Energy densities of nuclear fuels. Energy per fission does not include the energy lost to neutrinos since it is practically unrecoverable.

Material	Energy released per fission (MeV)	Atomic weight (g/mol)	Energy density (MJ/kg)
U-235	193.4	235.04	7979,390,000
U-238/Pu-239	198.9	238.05	80,620,000
Th-232/U-233	191.0	232.04	79,420,000

Complications

In a nuclear reactor, fission isn't the only process that releases energy. The actinides, fission products, and even structural and coolant nuclides often undergo capture reactions that release energy without fissioning. The fraction of energy released by a nuclear reactor by these reactions can be on the order of 10% of the total power of the reactor.

Nuclear Fuel Cycle

A nuclear fuel cycle is the path that we put heavy atoms through in order to extract energy from them, starting at the day we find them and ending when their wastes have decayed to stability and are no longer dangerous. Fuel cycles can take on a wide variety of configurations, and different configurations may make more sense than others in certain areas based on natural resource availability, energy growth projections, and politics. All commercial power-producing reactors in the USA are operating on a once-through cycle (which is more of a line than a cycle), while some in Europe and Asia go through a once- or twice-recycled cycle (which sounds funny). The economics, politics, and long-term sustainablity of nuclear energy depend critically on fuel cycles.

The Front End of the Fuel Cycle (From the Mine to the Reactor)

When whipping up a new fuel cycle, you usually have several processes to mix and match. These typically fall into three broad categories. First, we have the front end – steps used to prepare heavy atoms for insertion into nuclear reactors. So, you might survey the land, find uranium (or thorium) ore, dig it up, convert it to a gas so that you can enrich it, enrich it, convert it to a solid fuel form, and then fabricate it into fuel assemblies.

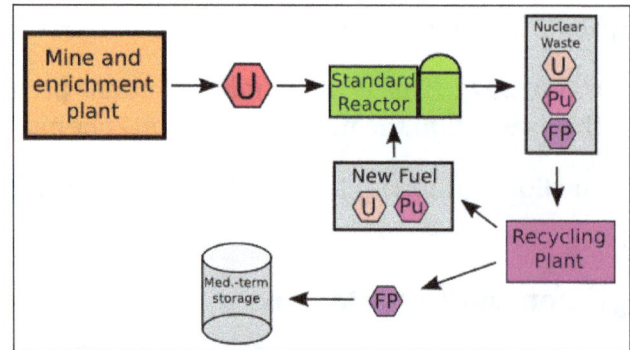

An example of a closed fuel cycle. Here nuclear material is recycled.

The In-reactor Part of the Fuel Cycle

The second category of fuel cycles is where we split atoms to generate energy. We can put fertile material around the reactor core and breed new fissile fuel, we can use liquid fuel with online re-fueling and fission-product separation, we can strategically place transuranic targets to reduce the toxicity of nuclear waste, we can perform alchemy, converting common atoms to rare or valuable ones, etc. We can reprocess, recycle and refabricate here. In reality, we are inhibited by economics, politics, licensing, and operations experience. Because of these, the most common type of fuel cycle has only one thing in this category: burn Uranium fuel rods in a reactor for about 3-5 years and remove them.

Nuclear Waste Disposal

Once we've finished getting energy out of the heavy atoms, we must dispose of the left-over ones and the dangerously radioactive minor actinides and fission products (the two smaller atoms created by splitting one large one). Processes to safely and responsibly accomplish this are the third category of our dynamic fuel cycle cast, called the back-end. The internationally accepted approach to waste disposal is to first ensure that the nuclides are immobilized (put in a material with low leach ability, good mechanical strength, and the capability to hold large amounts of waste) and then placed underground. Of course, arguments arise in how these two steps should be done. Materials used for immobilization are typically ceramics or glass. To protect against criticality accidents, these materials often have neutron-eating atoms such as boron mixed in. Finding a place to bury it is certainly a politically hot topic. The idea is to have as many barriers as is practical between the dangerous nuclides and the environment.

Reprocessing of Nuclear Fuel

Nuclear fuel reprocessing or nuclear recycling, involves the recovery of fissile material (plutonium and enriched uranium) and the separation of waste products from 'spent' (used) fuel rods from nuclear reactors.

Once reactor fuel (uranium or thorium) is used in a reactor, it can be treated and put into another reactor as fuel. In fact, typical reactors only extract a few percent of the energy in their fuel. You could power the entire US electricity grid off of the energy in nuclear waste for almost 100 years. If you recycle the waste, the final waste that is left over decays to harmlessness within a few hundred years, rather than a million years as with standard (unrecycled) nuclear waste.

Nuclear Transformations

Before you go on, recall that Uranium exists in nature as 2 isotopes: the less common U-235, and the more common U-238. Conventional reactors mainly split U-235 to produce power, and the U-238 is often considered useless. When a standard reactor runs low on U-235, it must be refueled, even though there is a lot of U-238 still in there.

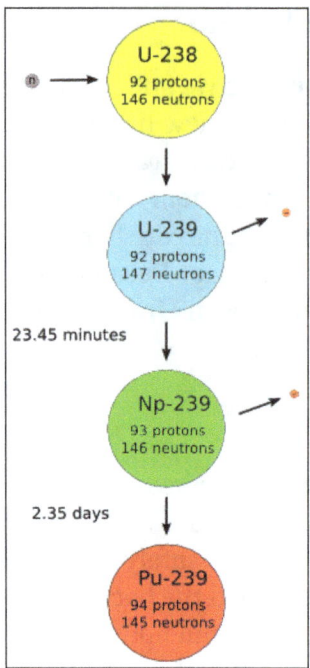

A common type of nuclear reaction is called beta-decay. When a nucleus has more neutrons than it would like to have, it often beta-decays by breaking a neutron into a proton and an electron. The electron (called a beta-particle in this case, since it originated in the nucleus) flies off into nature, and the main result seen in the nucleus is a neutron converting to a proton.

When U-238 absorbs a neutron in a nuclear reactor, it becomes U-239, which is just the isotope of Uranium with one extra neutron than U-238. This beta-decays quickly and becomes Np-239. Then, the Np-239 beta-decays again to become Pu-239, which is a fissile isotope that can power nuclear reactors.

The "useless" U-238 is the secret to recycling nuclear fuel. When it absorbs a single neutron, it goes through a series of nuclear reactions within a few days and turns into a very split table isotope of Plutonium, Pu-239. The Pu-239 acts a lot like the U-235 that powers conventional reactors, so if you convert your U-238 to Pu-239 as you run your reactor, you can then use that Pu-239 to continue powering your reactor, or others.

A nuclear fuel cycle is the path that nuclear fuel (Uranium, Thorium, Plutonium, etc.) takes as it is used to generate power in a nuclear reactor.

Once-through Cycle

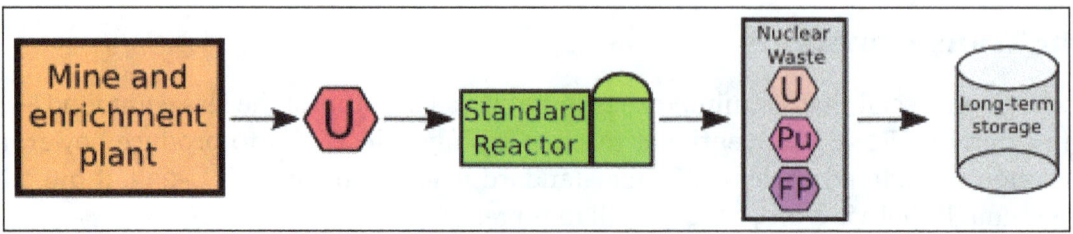

A once-through fuel cycle.

The simplest fuel cycle is the once-through cycle. It is the de-facto standard in most operating nuclear power plants, with a few exceptions in Europe and Asia. Uranium is mined, enriched, used in a reactor (where it becomes radioactive nuclear waste), and then stored until it is no longer dangerously radioactive. While this cycle is cheap, there are two major problems with it. Firstly, the waste is radioactive for hundreds of thousands of years. No one has been able to design a repository that is convincingly capable of storing material for that long. Secondly, Uranium is not the most abundant element on Earth, and in this kind of cycle, the global supply of cheap uranium could run low within 200 years. So much for sustainability. There are some deep-burn once-through cycles out there that have good sustainability properties though.

Closed Fuel Cycle

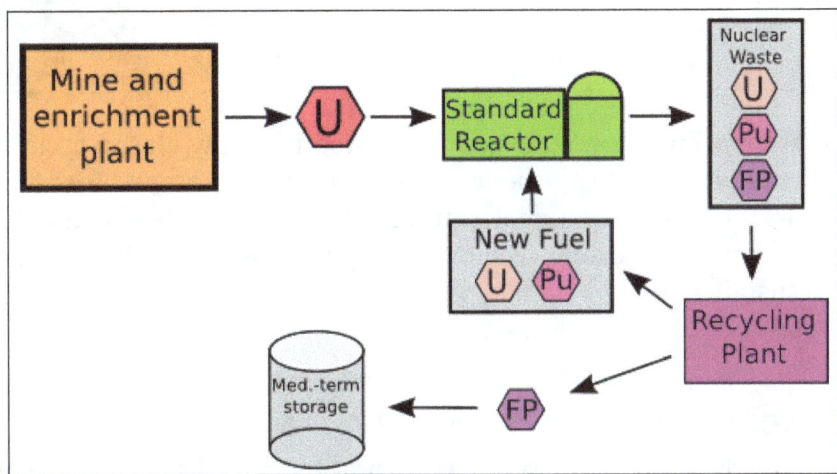

A closed fuel cycle.

Closing the fuel cycle involves recycling the nuclear waste as new fuel. Since the main component of nuclear waste is Uranium-238 (which can be transmuted to Plutonium, especially with advanced breeder reactors), we can get more energy out of the waste than in a once-through cycle. The recycling plant separates the good stuff from the bad stuff. The bad stuff is mostly fission products, the atoms that a Uranium atom becomes after it splits in the fission process. These fission products mostly decay to safe levels within 300-500 years, which is significantly shorter than standard nuclear waste. So, by closing the fuel cycle with standard reactors, we address the issue of nuclear waste identified in the once-through cycle. In this case, nuclear waste is a tractable problem. But most of the reactivity is coming from the mine, since standard reactors burn most of the fissile nuclide, U-235. Also, the reprocessing technology is expensive and separates out pure Plutonium, which could possibly be stolen, bringing a rogue entity closer to having a nuclear weapon. For these reasons, the USA does not currently recycle. There are ways to solve these issues.

Breeder Fuel Cycle

Breeder reactors can create as much or more fissile material (atoms that readily split) than they use. These special reactors are designed to have extra neutrons flying around, so that some can convert U-238 to Pu-239 and the others can run the reactor. Often, these special reactors are deemed "fast" reactors because the neutrons are moving through the reactor at higher speeds, on average. In a full breeder fuel cycle, we get the maximum use of the Uranium resources on Earth, and what we already know exists can last tens of thousands of years.

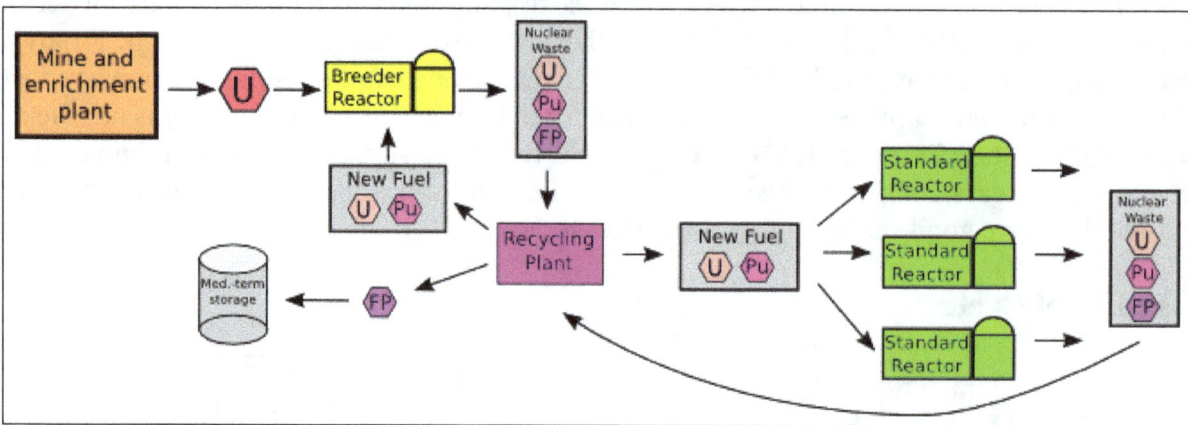

The cycle has cost and proliferation concerns associated with any closed cycle. Additionally, we have significantly less operational experience with breeder reactors, so we would need to train builders and operators for such a machine. Using a Thorium cycle instead of a Uranium-Plutonium cycle may allow breeding in less exotic reactors. Using this kind of fuel cycle, nuclear power can truly be considered sustainable.

References

- Nuclear-fuel: energyeducation.ca, Retrieved 16 June, 2019

- Fuel-fabrication, nuclear-fuel-cycle-conversion-enrichment-and-fabrication: world-nuclear.org, Retrieved 27 April, 2019

- Energy-density: whatisnuclear.com, Retrieved 16 June, 2019

- Fuel-cycle: whatisnuclear.com, Retrieved 09 January, 2019

- Recycling: whatisnuclear.com, Retrieved 26 March, 2019

6

Radioactive Waste Management

Radioactive waste refers to nuclear material which is left after a nuclear reaction. Radioactive waste management involves the treatment, storage, and disposal of liquid, airborne, and solid effluents from the nuclear industry's operations. The topics elaborated in this chapter will help in gaining a better perspective about radioactive waste management.

Radioactive Waste

Radioactive waste is nuclear fuel that is produced after being used inside of a nuclear reactor. Although it looks the same as it did before it went inside of the nuclear producer it has changed compounds and is nothing like the same. What is left is considered radioactive material and is very dangerous to anyone. This is very dangerous and remains this way for not just a few years but for thousands of years. It must be handled in the right manner so not to cause a ton of devastation in the world. It could take just seconds to die from exposure to radioactive materials. In short, radioactive waste is a kind of waste in gas, liquid or solid form that contain radioactive nuclear substance.

There are many industries like mining, defense, medicine, scientific research, nuclear power generation which produce by-products that include radioactive waste. The radioactive waste can remain radioactive for few months, years or even hundreds of years and the level of radioactivity can vary. The radioactive waste is extremely toxic as it can remain radioactive for so long and can cause acute radiation sickness when it first comes out of the reactor, if you stood within a few meters of it while it was unshielded.

Those who work inside of these facilities must be trained extensively to protect themselves and the rest of the world. It is very dangerous, although some types of radioactive waste are considered to be more harmful than other types. Since it is so hazardous and toxic, finding suitable disposal sites for radioactive waste remains a tedious task. Therefore, safe disposal site is required to ensure safety of humans, animals and this environment from toxic gases.

Types of Radioactive Waste

According to the United States Nuclear Regulatory Commission more than 104 licensed nuclear facilities are located inside of the United States. These reactors total 20% of the energy consumption being used. There are five types of radioactive waste- high level, low level, intermediate level, mining and milling and transuranic waste. All types of nuclear wastes have their own separate storage and clean-up procedures.

High-level Waste

You will find that there are two types of nuclear reactors. These types are the pressurized and boiler water reactors. High-level nuclear waste, simply put, is spent fuel that is still present after it has been used inside of nuclear reactors. This radioactive waste has to cool off for several years and is considered to be very dangerous. The cooling process of this waste usually takes place inside of deep pools of water that are several hundred feet deep. These pools can be located on-site of off-site of the nuclear facility although the off-site facilities are limited and must be approved by the EPA .

This type of waste is hazardous to people for many reasons, but especially because it remains radioactive. High level waste accounts for 95% of the total radioactivity produced in the nuclear reactor. This type of nuclear waste is very dangerous. It must consistently go through a process to keep it cool and the radioactive material under control. High level waste can have short and long lived components depending upon the time it will take for the radioactivity to decrease to levels that is not considered harmful for humans and surrounding environment.

Intermediate-level Waste

Intermediate-level waste contains high amount of radioactivity than low-level and less than high-level. This type of waste typically requires shielding during handling and interim storage. This type of waste typically includes refurbishment waste, ion-exchange resins, chemical sludges and metal fuel cladding. The intermediate level waste contains 4% of the entire radioactivity. Intermediate-level waste that requires long term management is transferred to an authorized waste management operator.

Low-level Waste

Most of the radioactive waste that is around today is considered to be low level. In fact, about 90% of all nuclear waste is low level. Nuclear reactors, hospitals, dental offices, and similar types of facilities often use low-level nuclear waste materials on a daily basis and it is needed in order to provide the services that are offered within these facilities. Low-level nuclear waste is not dangerous, and any of it can be disposed of inside of a landfill. This is the reason why it does not require shielding during handling and transport.

Even so there is a strict criterion in which it must be handled and disposed of. Without the proper disposal what is not dangerous has the possibilities of becoming that way. This is not a chance that you should be willing to take when it is so very easy to protect yourself. The low level waste contains just 1% of the radioactivity of all radioactive waste.

Mining and Milling

Tailings and waste rock are generates by mining and milling of uranium ore. The tailings materials is covered with water and have the consistency of fine sand, when dried. It is produced by grinding the ore and the chemical concentration of uranium. After few months, the tailings material contains 75% of the radioactivity of the original ore.

Clean and mineralized waste rock is produced during mining activities which must be excavated to access to access uranium ore body. It has little or no concentration of uranium. While clean waste rock can be used for construction purposes mineralized waste rock could generate acid when left on the surfaced indefinitely that could affect surrounding environment.

Transuranic Waste

Transuranic waste, or TRU waste contains more than 3700 be per gram of elements. It is much heavier than uranium. This type of waste is produced through nuclear waste reprocessing procedures in most cases. This is one of the least worried about types of radioactive waste that is out there but it is worth mentioning since it is a part of nuclear waste.

Other Classifications

Non-commercial activities can bring nuclear waste to the forefront. There are several activities, including by-product materials and uranium mining, to name a few. The clean-up standards and procedures for these activities vary and are set forth by the Environmental Protection Agency.

Now you have the knowledge of radioactive waste and how it is disposed of in a safe manner.

Radioactive Waste Management

Radioactive waste management involves the treatment, storage, and disposal of liquid, airborne, and solid effluents from the nuclear industry's operations.

Site of Waste Production

Radioactive waste is produced at all stages of the nuclear fuel cycle – the process of producing electricity from nuclear materials. The fuel cycle involves the mining and milling of uranium ore, its processing and fabrication into nuclear fuel, its use in the reactor, its reprocessing (if conducted), the treatment of the used fuel taken from the reactor, and finally, disposal of the waste. Whilst waste is produced during mining and milling and fuel fabrication, the majority comes from the actual 'burning' of uranium to produce electricity. Where the used fuel is reprocessed, the amount of waste is reduced materially.

Mining through to Fuel Fabrication

Traditional uranium mining generates fine sandy tailings, which contain virtually all the naturally occurring radioactive elements found in uranium ore. The tailings are collected in engineered dams and finally covered with a layer of clay and rock to inhibit the leakage of radon gas, and to ensure long-term stability. In the short term, the tailings material is often covered with water. After a few months, the tailings material contains about 75% of the radioactivity of the original ore. Strictly speaking these are not classified as radioactive waste.

Uranium oxide concentrate from mining, essentially 'yellowcake' (U_3O_8), is not significantly radioactive – barely more so than the granite used in buildings. It is refined then converted to uranium hexafluoride (UF_6) gas. As a gas, it undergoes enrichment to increase the U-235 content from 0.7% to about 3.5%. It is then turned into a hard ceramic oxide (UO_2) for assembly as reactor fuel elements.

The main by-product of enrichment is depleted uranium (DU), principally the U-238 isotope, which is stored either as UF_6 or U_3O_8. Some DU is used in applications where its extremely high density makes it valuable, such as for the keels of yachts and military projectiles. It is also used (with reprocessed plutonium) for making mixed oxide (MOX) fuel and to dilute highly-enriched uranium from dismantled weapons, which can then be used for reactor fuel.

Electricity Generation

In terms of radioactivity, the major source arising from the use of nuclear reactors to generate electricity comes from the material classified as HLW. Highly radioactive fission products and transuranic elements are produced from uranium and plutonium during reactor operations, and are contained within the used fuel. Where countries have adopted a closed cycle and reprocess used fuel, the fission products and minor actinides are separated from uranium and plutonium and treated as HLW. In countries where used fuel is not reprocessed, the used fuel itself is considered a waste and therefore classified as HLW.

LLW and ILW is produced as a result of general operations, such as the cleaning of reactor cooling systems and fuel storage ponds, and the decontamination of equipment, filters, and metal components that have become radioactive as a result of their use in or near the reactor.

Reprocessing of used Fuel

Any used fuel will still contain some of the original U-235 as well as various plutonium isotopes which have been formed inside the reactor core, and U-238. In total these account for some 96% of the original uranium and over half of the original energy content (ignoring U-238). Used nuclear fuel has long been reprocessed to extract fissile materials for recycling and to reduce the volume of HLW. Several European countries, as well as Russia, China, and Japan have policies to reprocess used nuclear fuel.

Reprocessing allows for a significant amount of plutonium to be recovered from used fuel, which is then mixed with depleted uranium oxide in a MOX fabrication plant to make fresh fuel. This process allows some 25-30% more energy to be extracted from the original uranium ore, and significantly reduces the volume of HLW (by about 85%). The IAEA estimates that of the 370,000 metric tonnes of heavy metal (MTHM) produced since the advent of civil nuclear power production, 120,000 MTHM has been reprocessed. In addition, the remaining HLW is significantly less radioactive – decaying to the same level as the original ore within 9000 years (vs. 300,000 years).

Commercial reprocessing plants currently operate in France, the UK, and Russia. Another is being commissioned in Japan, and China plans to construct one too. France undertakes reprocessing for utilities in other countries, and a lot of Japan's fuel has been reprocessed there, with both waste and recycled plutonium in MOX fuel being returned to Japan.

The main historical and current process is Purex, a hydrometallurgical process. The main prospective ones are electrometallurgical – often called pyro-processing since it happens to be hot. With it, all actinide anions (notably uranium and plutonium) are recovered together. Whilst not yet operational, these technologies will result in waste that only needs 300 years to reach the same level of radioactivity as the originally mined ore.

Storage pond for used fuel at the Thermal Oxide Reprocessing Plant.

Decommissioning Nuclear Plants

In the case of nuclear reactors, about 99% of the radioactivity is associated with the fuel. Apart from any surface contamination of plant, the remaining radioactivity comes from 'activation products' such as steel components which have long been exposed to neutron irradiation. Their atoms are changed into different isotopes such as iron-55, cobalt-60, nickel-63, and carbon-14. The first two are highly radioactive, emitting gamma rays, but with correspondingly short half-lives so that after 50 years from final shutdown their hazard is much diminished. Some caesium-137 may also be found in decommissioning wastes.

Some scrap material from decommissioning may be recycled, but for uses outside the industry very low clearance levels are applied, so most is buried and some is recycled within the industry.

Legacy Waste

In addition to the routine waste from current nuclear power generation there is other radioactive waste referred to as 'legacy waste'. This waste exists in several countries that pioneered nuclear power and especially where power programs were developed out of military programs. It is sometimes voluminous and difficult to manage, and arose in the course of those countries getting to a position where nuclear technology is a commercial proposition for power generation. It represents a liability which is not covered by current funding arrangements. In the UK, some £73 billion (undiscounted) is estimated to be involved in addressing this waste – principally from Magnox and some early AGR developments – and about 30% of the total is attributable to military programs. In the USA, Russia, and France the liabilities are also considerable.

Non-nuclear Power Waste

In recent years, in both the radiological protection and radioactive waste management communities, there has been increased attention on how to effectively manage nonpower related nuclear waste. All countries, including those that do not have nuclear power plants, have to manage radioactive waste generated by activities unrelated to the production of nuclear energy, including: national laboratory and university research activities; used and lost industrial gauges and radiography sources; and nuclear medicine activities at hospitals. Although much of this waste is not long-lived, the variety of the sources makes any general assessment of physical or radiological characteristics difficult. The relatively source-specific nature of the waste poses questions and challenges for its management at a national level.

Treatment and Conditioning

Treatment involves operations intended to change waste streams' characteristics to improve safety or economy. Treatment techniques may involve compaction to reduce volume, filtration or ion exchange to remove radionuclide content, or precipitation to induce changes in composition.

Conditioning is undertaken to change waste into a form that is suitable for safe handling, transportation, storage, and disposal. This step typically involves the immobilisation of waste in containers. Liquid LLW and ILW are typically solidified in cement, whilst HLW is calcined/dried then vitrified in a glass matrix. Immobilised waste will be placed in a container suitable for its characteristics.

Storage and Disposal

Storage of waste may take place at any stage during the management process. Storage involves maintaining the waste in a manner such that it is retrievable, whilst ensuring it is isolated from the external environment. Waste may be stored to make the next stage of management easier (for example, by allowing its natural radioactivity to decay). Storage facilities are commonly onsite at the power plant, but may be also be separate from the facility where it was produced.

Disposal of waste takes place when there is no further foreseeable use for it, and in the case of HLW, when radioactivity has decayed to relatively low levels after about 40-50 years.

LLW and Short-lived ILW

Most LLW and short-lived ILW are typically sent to land-based disposal immediately following packaging. This means that for the majority (>90% by volume) of all of the waste types, a satisfactory disposal means has been developed and is being implemented around the world.

Near-surface disposal facilities are currently in operation in many countries, including:

- UK: LLW Repository at Drigg in Cumbria operated by UK Nuclear Waste. Management (a consortium led by Washington Group International with Studsvik UK, Serco, and Areva) on behalf of the Nuclear Decommissioning Authority.

- Spain: El Cabril LLW and ILW disposal facility operated by ENRESA.

- France: Centre de l'Aube and Morvilliers operated by ANDRA.

- Sweden: SFR at Forsmark operated by SKB.

- Finland: Olkiluoto and Loviisa, operated by TVO and Fortum.

- Russia: Ozersk, Tomsk, Novouralsk, Sosnovy Bor, operated by NO RAO.

- South Korea: Wolseong, operated by KORAD.

- Japan: LLW Disposal Center at Rokkasho-Mura operated by Japan Nuclear Fuel Limited.

- USA: five LLW disposal facilities: Texas Compact facility near the New Mexico border, operated by Waste Control Specialists; Barnwell, South Carolina; Clive, Utah; Oak Ridge, Tennessee – all operated by Energy Solutions; and Richland, Washington – operated by American Ecology Corporation.

Some low-level liquid waste from reprocessing plants is discharged to the sea. This includes radionuclides which are distinctive, notably technetium-99 (sometimes used as a tracer in environmental studies), and this can be discerned many hundred kilometres away. However, such discharges are regulated and controlled, and the maximum radiation dose anyone receives from them is a small fraction of natural background radiation.

Nuclear power stations and reprocessing plants release small quantities of radioactive gases (*e.g.* krypton-85 and xenon-133) and trace amounts of iodine-131 to the atmosphere. However, krypton-85 and xenon-133 are chemically inert, all three gases have short half-lives, and the radioactivity in the emissions is diminished by delaying their release. The net effect is too small to warrant consideration in any life-cycle analysis. A little tritium is also produced but regulators do not consider its release to be significant.

Long-lived ILW and HLW

The long timescales over which some ILW and HLW – including used fuel when considered a waste – remains radioactive has led to universal acceptance of the concept of deep geological disposal. Many other long-term waste management options have been investigated, but deep disposal in a mined repository is now the preferred option in most countries. The Waste Isolation Pilot Plant

(WIPP) deep geological waste repository is in operation in the US for the disposal of transuranic waste – long-lived ILW from military sources, contaminated with plutonium.

To date there has been no practical need for final HLW repositories. The used fuel may either by reprocessed or disposed of directly. Either way, there is a strong technical incentive to delay final disposal of HLW for about 40-50 years after removal, at which point the heat and radioactivity will have reduced by over 99%. Interim storage of used fuel is mostly in ponds associated with individual reactors, or in a common pool at multi-reactor sites, or occasionally at a central site. At present there is about 250,000 tonnes of used fuel in storage. Over two-thirds of this is in storage ponds, with an increasing proportion in dry storage.

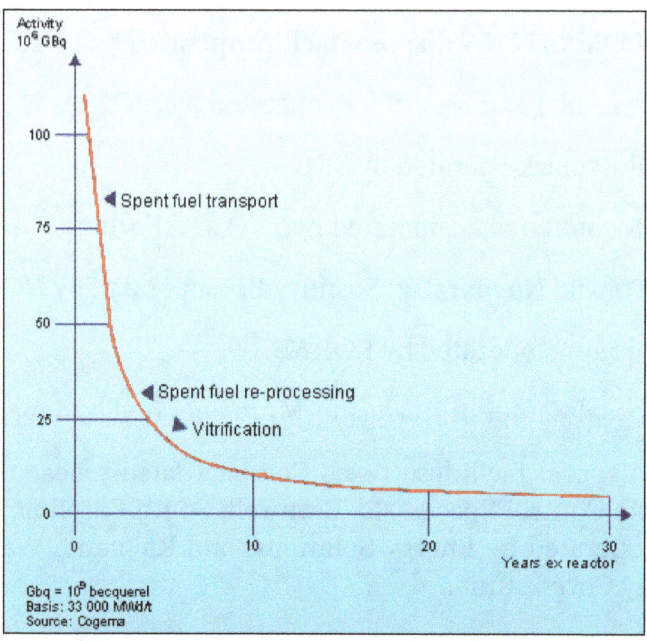

Illustrative decay in radioactivity of fission products – one tonne of spent PWR fuel.

Storage ponds at reactors, and those at centralised facilities such as CLAB in Sweden, are 7-12 metres deep to allow for several metres of water over the used fuel (assembled in racks typically about 4 metres long and standing on end). The multiple racks are made of metal with neutron absorbers incorporated. The circulating water both shields and cools the fuel. These pools are robust constructions made of thick reinforced concrete with steel liners. Ponds at reactors are often designed to hold all the used fuel produced over the planned operating lifetime of the reactor.

Water-filled storage pools at the Central Interim Storage Facility for Spent Nuclear Fuel (CLAB) facility in Sweden.

Some fuel that has cooled in ponds for at least five years is stored in dry casks or vaults with air circulation inside concrete shielding. One common system is for sealed steel casks or multi-purpose canisters (MPCs) each holding up to about 40 fuel assemblies with inert gas. Casks/MPCs may also be used for the transport and eventual disposal of the used fuel. For storage, each is enclosed in a ventilated storage module made of concrete and steel. These are commonly standing on the surface, about 6 m high, and cooled by air convection, or they may be below grade, with just the tops showing. The modules are robust and provide full shielding. Each cask has up to 45 kW heat load.

If used reactor fuel is reprocessed, the resulting liquid HLW must be solidified. The HLW also generates a considerable amount of heat and requires cooling. It is vitrified into borosilicate (Pyrex) glass, sealed into heavy stainless steel cylinders about 1.3 metres high, and stored for eventual disposal deep underground. This material has no conceivable future use and is universally classified as waste. France has two commercial plants to vitrify HLW left over from reprocessing fuel, and there are also plants active in the UK and Belgium. The capacity of these western European plants are 2,500 canisters (1000 t) a year, and some have been operating for three decades. By mid-2009, the vitrification plant at Sellafield, UK, had produced its 5000th canister of vitrified HLW, representing 3000 m^3 of liquor reduced to 750 m^3 of glass. The plant currently fills about 400 canisters per year.

The Australian Synroc (synthetic rock) system is a more sophisticated way to immobilise such waste, and this process may eventually come into commercial use for civil wastes.

If used reactor fuel is not reprocessed, it will still contain all the highly radioactive isotopes. Spent fuel that is not reprocessed is treated as HLW for direct disposal. It too generates a lot of heat and requires cooling. However, since it largely consists of uranium (with a little plutonium), it represents a potentially valuable resource, and there is an increasing reluctance to dispose of it irretrievably.

For final disposal, to ensure that no significant environmental releases occur over tens of thousands of years, 'multiple barrier' geological disposal is planned. This technique will immobilise the radioactive elements in HLW and long-lived ILW, and isolate them from the biosphere. The multiple barriers are:

- Immobilisation of waste in an insoluble matrix such as borosilicate glass or synthetic rock (fuel pellets are already a very stable ceramic, UO_2).

- Contain waste sealed inside a corrosion-resistant container, such as stainless steel.

- Isolate waste from people and the environment, so eventually locate it deep underground in a stable rock structure.

- Delay any significant migration of radionuclides from the repository, so surround containers with an impermeable backfill such as bentonite clay if the repository is wet.

Due to the long-term nature of these management plans, sustainable options must have one or more pre-defined milestones where a decision could be taken on which option to proceed with.

A current question is whether waste should be emplaced so that it is readily retrievable from repositories. There are sound reasons for keeping such options open – in particular, it is possible that future generations might consider the buried waste to be a valuable resource. On the other hand, permanent closure might increase long-term security of the facility. After being buried for about 1,000 years most of the radioactivity will have decayed. The amount of radioactivity then remaining would be similar to that of the naturally-occurring uranium ore from which it originated, though it would be more concentrated. In mined repositories, which represent the main concept being pursued, retrievability can be straightforward, but any deep borehole disposal is permanent.

Loading silos with canisters containing vitrified HLW in the UK.
Each disc on the floor covers a silo holding ten canisters.

France's 2006 waste law says that HLW disposal must be 'reversible', which was clarified in a 2015 amendment to mean guaranteeing long-term flexibility in disposal policy, while 'retrievable' referred to short-term practicality. France, Switzerland, Canada, Japan, and the USA require retrievability. That policy is followed also in most other countries, though this presupposes that in the long-term, the repository would be sealed to satisfy safety requirements.

Natural Precedents for Geological Disposal

Nature has already proven that geological isolation is possible through several natural examples (or 'analogues'). The most significant case occurred almost 2 billion years ago at Oklo, in what is now Gabon in West Africa, where several spontaneous nuclear reactors operated within a rich vein of uranium ore. (At that time the concentration of U-235 in all natural uranium was about 3%.) These natural nuclear reactors continued for about 500,000 years before dying away. They produced all the radionuclides found in HLW, including over 5 tonnes of fission products and 1.5 tonnes of plutonium, all of which remained at the site and eventually decayed into non-radioactive elements.

The study of such natural phenomena is important for any assessment of geologic repositories, and is the subject of several international research projects.

Funding Waste Management

Nuclear power is the only large-scale energy-producing technology that takes full responsibility for all its waste and fully costs this into the product. Financial provisions are made for managing all kinds of civilian radioactive waste. The cost of managing and disposing of nuclear power plant waste typically represents about 5% of the total cost of the electricity generated.

Most nuclear utilities are required by governments to put aside a levy (e.g. 0.1 cents per kilowatt hour in the USA, 0.14 ¢/kWh in France) to provide for the management and disposal of their waste.

The actual arrangements for paying for waste management and decommissioning vary. The key objective is, however, always the same: to ensure that sufficient funds are available when they are needed. There are three main approaches:

- Provisions on the balance sheet: Sums to cover the anticipated cost of waste management and decommissioning are included on the generating company's balance sheet as a liability. As waste management and decommissioning work proceeds, the company has to ensure that it has sufficient investments and cashflow to meet the required payments.

- Internal fund: Payments are made over the operating lifetime of the nuclear facility into a special fund that is held and administered within the company. The rules for the management of the fund vary, but many countries allow the fund to be re-invested in the assets of the company, subject to adequate securities and investment returns.

- External fund: Payments are made into a fund that is held outside the company, often within government or administered by a group of independent trustees. Again, rules for the management of the fund vary. Some countries only allow the fund to be used for waste management and decommissioning purposes, whilst others allow companies to borrow a percentage of the fund to reinvest in their business.

How much Waste is Produced?

The volume of high-level radioactive waste (HLW) produced by the civil nuclear industry is small. The IAEA estimates that 370,000 tonnes of heavy metal (tHM) in the form of used fuel have been discharged since the first nuclear power plants commenced operation. Of this, the agency estimates that 120,000 tHM have been reprocessed. The IAEA estimates that the disposal volume of the current solid HLW inventory is approximately 22,000m³. For context, this is a volume roughly equivalent to a three metre tall building covering an area the size of a soccer pitch.

Disposal volumes vary based on the chosen solution for waste disposal. In arriving at its estimate, the IAEA has made assumptions with respect to packaging and repository design for countries without confirmed disposal solutions based on the plans proposed by countries more advanced in the process.

The amounts of ILW, LLW, and VLLW produced are greater in volume, but are much less radioactive. Given its lower inherent radioactivity, the majority of waste produced by nuclear power production and classified as LLW or VLLW has already been placed in disposal. The IAEA estimates that over 80% of all LLW and VLLW produced to date are in disposal. For ILW, the agency estimates that about 20% is in disposal, with the balance in storage.

Table: Nuclear waste inventory (IAEA estimates, 2018).

	Solid radioactive waste in storage (m³)	Solid radioactive waste in disposal (m³)	Proportion of waste type in disposal
VLLW	2,356,000	7,906,000	77%
LLW	3,479,000	20,451,000	85%
ILW	460,000	107,000	19%
HLW	22,000	0	0%

All hazardous waste requires careful management and disposal, not just radioactive waste. The amount of waste produced by the nuclear power industry is small relative to both other forms of electricity generation and general industrial activity. For example, in the UK – the world's oldest nuclear industry – the total amount of radioactive waste produced to date, and forecast to 2125, is about 4.9 million tonnes. After all waste has been packaged, it is estimated that the final volume would occupy a space similar to that of a large, modern soccer stadium. This compares with an annual generation of 200 million tonnes of conventional waste, of which 4.3 million tonnes is classified as hazardous. About 94% of radioactive waste in the UK is classified as LLW, about 6% is ILW, and less than 0.03% is classified as HLW.

In over 50 years of civil nuclear power experience, the management and disposal of civil nuclear waste has not caused any serious health or environmental problems, nor posed any real risk to the general public. Alternatives for power generation are not without challenges, and their undesirable by-products are generally not well controlled.

To put the production and management of nuclear waste in context, it is important to consider the non-desirable by-products – most notably carbon dioxide emissions – of other large-scale commercial electricity generating technologies. In 2016, nuclear power plants supplied 2,417 TWh of electricity, 11% of the world's total consumption. Fossil fuels supplied 67%, of which coal contributed the most (8,726 TWh), followed by gas (4,933 TWh), and oil (1,068 TWh). If the 11% of electricity supplied by nuclear power had been replaced by gas – by far the cleanest burning fossil fuel – an additional 2,388 million tonnes of CO_2 would have been released into the atmosphere; the equivalent of putting an additional 250 million cars on the road.

Table: CO_2 emissions avoided through the use of nuclear power.

	Lifecycle emissions (gCO_2eq/kWh)	Estimated emissions to produce 2417 TWh electricity (million tonnes CO_2)	Potential emissions avoided through use of nuclear power (million tonnes CO_2)	Potential emissions avoided through use of nuclear (million cars equivalent)
Nuclear Power	12	29	NA	NA
Gas (CCS)	490	1184	1155	250
Coal	820	1981	1952	400

In addition to producing very significant emissions of carbon, hydrocarbon industries also create significant amounts of radioactive waste. The radioactive material produced as a waste product from the oil and gas industry is referred to as 'technologically enhanced naturally occurring radioactive materials' (Tenorm). In oil and gas production, radium-226, radium-228, and lead-210 are deposited as scale in pipes and equipment in many parts of the world. Published data show radionuclide concentrations in scales up to 300,000 Bq/kg for Pb-210, 250,000 Bq/kg for Ra-226, and 100,000 Bq/kg for Ra-228. This level is 1000 times higher than the clearance level for recycled material (both steel and concrete) from the nuclear industry, where anything above 500 Bq/kg may not be cleared from regulatory control for recycling.

The largest Tenorm waste stream is coal ash, with around 280 million tonnes arising globally each year, carrying uranium-238 and all its non-gaseous decay products, as well as thorium-232 and its progeny. This ash is usually just buried, or may be used as a constituent in building materials. As such, the same radionuclide, at the same concentration, may be sent to deep disposal if from the nuclear industry, or released for use in building materials if in the form of fly ash from the coal industry.

7

Nuclear Safety and Security

Nuclear safety deals with rules of conduct for the safety of the workers and nuclear security refers to the prevention and detection of theft, inappropriate access, damage, etc. to the nuclear facility. Some of the fields that fall under its domain are nuclear accidents, nuclear safety, nuclear criticality safety, nuclear reactor safety, nuclear power plant safety systems, etc. This chapter has been carefully written to provide an easy understanding of these fields which associates with nuclear safety and security.

Nuclear Accidents

Of all the environmental disaster events that humans are capable of causing, nuclear disasters have the greatest damage potential. The radiation release associated with a nuclear disaster poses significant acute and chronic risks in the immediate environs and chronic risk over a wide geographic area. Radioactive contamination, which typically becomes airborne, is long-lived, with half-lives guaranteeing contamination for hundreds of years.

Concerns over potential nuclear disasters center on nuclear reactors, typically those used to generate electric power. Other concerns involve the transport of nuclear waste and the temporary storage of spent radioactive fuel at nuclear power plants. The fear that terrorists would target a radiation source or create a "dirty bomb" capable of dispersing radiation over a populated area was added to these concerns following the 2001 terrorist attacks on New York City and Washington, D.C.

Radioactive emissions of particular concern include strontium-90 and cesium-137, both having thirty-year-plus half-lives, and iodine-131, having a short half-life of eight days but known to cause thyroid cancer. In addition to being highly radioactive, cesium-137 is mistaken for potassium by living organisms. This means that it is passed on up the food chain and bioaccumulated by that process. Strontium-90 mimics the properties of calcium and is deposited in bones where it may either cause cancer or damage bone marrow cells.

The Chernobyl Disaster

Concern became reality at 1:23 a.m. on April 25, 1986, when the worst civil nuclear catastrophe in history occurred at the nuclear power plant at Chernobyl, Soviet Union (which is now in Ukraine). More than thirty people were killed immediately. The radiation release was thirty to forty times that of the atomic bombs dropped on Japan during World War II. Hundreds of thousands of people were ultimately evacuated from the most heavily contaminated zone surrounding Chernobyl. Radiation spread to encompass almost all of Europe and Asia Minor; the world first learned of the disaster when a nuclear facility in Sweden recorded abnormal radiation levels.

Chernobyl had four RBMK-type reactors. These reactors suffer from instability at low power and are susceptible to rapid, difficult-to-control power increases. The accident occurred as workers were testing reactor number four. The test was being conducted improperly; as few as six control rods were in place despite orders stating that a minimum of thirty rods were necessary to maintain control, and the reactor's emergency cooling system had been shut down as part of the test. An operator error caused the reactor's power to drop below specified levels, setting off a catastrophic power surge that caused fuel rods to rupture, triggering explosions that first destroyed the reactor core and then blew apart the reactors' massive steel and concrete containment structure.

The health impacts of the Chernobyl explosion will never be fully known. It is estimated that some three million people still live in contaminated areas and almost ten thousand people still live in Chernobyl itself. The plant itself was not fully shut down until nearly fifteen years after the disaster. Studies by the Belarus Ministry of Health, located approximately eighty miles south of Chernobyl, found that rates of thyroid cancer began to soar in contaminated regions in 1990, four years after the radiation release. Gomel, Belarus, the most highly contaminated region studied, reported thirty-eight cases in 1991. Gomel normally recorded only one to two cases per year. Health officials in Turkey, 930 miles to the south, reported that leukemia rates are twelve times higher than before the Chenobyl accident.

Three Mile Island

The thriller China Syndrome, which warned that a nuclear power plant meltdown would blow a hole through the earth all the way to China and "render an area the size of Pennsylvania permanently uninhabitable" had been playing for eleven days when, at 4:00 am on March 28, 1979, Reactor 2 at the Three Mile Island (TMI) nuclear power plant suffered a partial meltdown. The plant was just downriver from Harrisburg, Pennsylvania.

Film story, reality, and perception all interplayed to create near national panic. The accident occurred sequentially. A minor problem caused the temperature of the primary coolant to rise. In one second, the reactor shut down but a relief valve that was supposed to close after ten seconds remained open. Plant instrumentation showed operators that a "close valve" signal had been sent. There was no instrumentation to tell them the valve itself was still open. The reactor's primary coolant drained away and the reactor core suffered serious damage. Fuel rods were damaged, leaking radioactive material into the cooling water and a high temperature chemical reaction created bubbles of hydrogen gas. One of these bubbles burned, creating fears that a larger hydrogen bubble would explode, possibly breaching the plant's containment structure. Some gases were purposefully vented into the atmosphere.

It took nearly a full month the bring the reactor into "cold shutdown" status. That said, there was never danger of a massive explosion and hundreds of readings taken by the Pennsylvania Department of Environmental Resources showed almost no iodine, and all readings were far below health limits. There was, however, widespread panic including an unordered mass evacuation. The greatest problem at TMI was a total failure of communication. Internal radioactivity levels, for example, were reported as ambient (outdoor) air readings.

The many health studies following TMI showed no evidence of abnormal cancer rates. For eighteen years, the Pennsylvania Department of Health maintained a registry of 30,000 people who lived within five miles of TMI; it found no evidence on unusual health trends. TMI's only health effect was psychological stress related to the accident.

While there were few long-term health effects, there is no doubt that the accident at TMI permanently changed both the nuclear industry and the Nuclear Regulatory Commission (NRC). "Public fear and distrust increased," the NRC notes in a fact sheet on TMI, "Regulations and oversight became broader and more robust, and management of the plants was scrutinized more carefully."

Nuclear Submarines

On August 12, 2000, an explosion in a torpedo tube sank the giant Russian nuclear submarine Kursk and its crew of 118 in the Barents Sea. Russian officials described the sinking as a "catastrophe that developed at lightning speed." A week later, divers opened the rear hatch of the sub but found no survivors. It took salvagers two years, but the Kursk and her two nuclear reactors were raised.

The Kursk was the sixth nuclear submarine to have sunk since 1963. The others all came to rest on the ocean floor at depths of more than 4,500 feet, far below where most marine life lives. They include two former Soviet submarines—one that sank east of Bermuda in 1986 and another that went down in the Bay of Biscay in 1970—and two U.S. nuclear submarines—the U.S.S. Thresher and U.S.S. Scorpion —which sank in the 1960s at the height of the Cold War.

U.S. Navy officials report there is little likelihood of radioactive release from the U.S. ships. Reactor fuel elements in American submarines are made of materials that are extremely corrosion resistant, even in sea water. The protective cladding on the fuel elements corrodes only a few millionths of an inch per year, meaning the reactor core could remain submerged in sea water for centuries without releases of fission products while the radioactivity decays.

Comprehensive deep ocean radiological monitoring operations were conducted at the Thresher site in 1965, 1977, 1983, and again in 1986. None of the samples obtained showed any evidence of release of radioactivity from the reactor fuel elements.

Nuclear Safety

Nuclear Safety is Safety applied to all activities involving radioactive material or ionizing radiation, notably:

- All types of nuclear reactors (Nuclear Power Plants — NPP, research reactors, propulsion reactors, etc.);

- Nuclear fuel cycle facilities (upstream and downstream);

- Hot laboratories and particle accelerators;

- Nuclear waste storage units;

- All types of nuclear facilities in decommissioning stage;

- All types of transport of radioactive material;

- Activities (including medical) involving radioactive sources.

Thus, from the Safety Glossary of the IAEA (International Atomic Energy Agency), Nuclear Safety is the achievement of proper operating conditions, prevention of accidents or mitigation of accident consequences, resulting in protection of workers, the public and the environment from undue radiation hazards.

In nuclear industry, safety is distinguished from two other important and complementary concepts:

- Radiation Protection;

- Safeguards of nuclear material (protection against nuclear proliferation).

Nuclear Safety General Objective

The fundamental safety objective is to protect people and the environment from harmful effects of ionizing radiation.

To ensure that facilities are operated and activities conducted so as to achieve the highest standards of safety that can reasonably be achieved, measures have to be taken to:

- Control the radiation exposure of people and the release of radioactive material to the environment;

- Restrict the likelihood of events that might lead to a loss of control over a nuclear reactor core, nuclear chain reaction, radioactive source or any other source of radiation;

- Mitigate the consequences of such events if they were to occur.

The fundamental safety objective applies for all facilities and activities and for all stages over the lifetime of a facility or radiation source, including planning, siting, design, manufacturing, construction, commissioning and operation, as well as decommissioning and closure.

This includes the associated transport of radioactive material and management of radioactive waste.

Eventually, it applies to all operating conditions, from normal operation to the management of radiological emergencies.

Thus, nuclear consulting companies need to consider safety for all nuclear power consulting.

Defence in Depth

Defence in Depth is originally a military concept.

Against a given threat, since no single defence is infallible, the idea of Defence in Depth is to stack several independent levels of defence.

Each level is designed to prevent so far as possible:

- Proceeding to the next level of defence;

- The consequences of previous level failure.

The International Nuclear Safety Group (INSAG) of the IAEA has settled Defence in Depth for Nuclear Safety in its report Defence in Depth in Nuclear Safety.

Defence in Depth in Nuclear Safety, thus structured in five levels, applies to all stages of design and operation of nuclear reactors and facilities.

The training courses given by SureDyna include a complete presentation of Defence in Depth as well as its applications to design and operation of nuclear reactors and nuclear facilities.

Meanwhile, three basic levels of Defence in Depth should be remembered:

- Prevention;
- Protection;
- Mitigation of consequences.

Example of Application

The best-known application of the Defence in Depth concept is the three containment barriers of radioactive materials around the core of a nuclear reactor:

- Fuel cladding;
- Reactor Primary Coolant boundary;
- Containment building.

However, although this example of defence in depth is excellent and rightly the most cited, it must still be considered as an instance of application of the concept of Defence depth, not as Defense in Depth itself.

Indeed, the barriers do not match the levels of Defence in Depth and should not be mistaken for them.

General Application

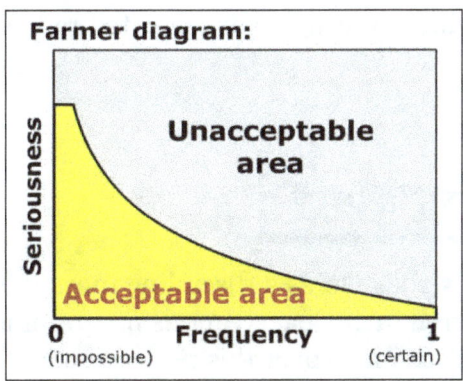

According to the different levels, the Defence in Depth approach defines categories of normal or abnormal operation situations (incidents and accidents) with associated means of design and operation (redundant equipment in case of failure, backup devices, specific procedures, etc.).

The requirements with regard to a given situation are all the more strong as this situation is likely to occur more frequently, as shown on the so-called "Farmer" diagram.

In any case, the levels of Défense in Depth must be independent in order to respect the Single Failure Criterion:

> A single event (incident) of a given category shall not generate an event of a higher category without other events occurring independently.

Safety Principles

- The prime responsibility for safety must rest with the person or organization responsible for facilities and activities that give rise to radiation risks.

- An effective legal and governmental framework for safety, including an independent regulatory body, must be established and sustained.

- Effective leadership and management for safety must be established and sustained in organizations concerned with, and facilities and activities that give rise to, radiation risks.

- Facilities and activities that give rise to radiation risks must yield an overall benefit.

- Protection must be optimized to provide the highest level of safety that can reasonably be achieved.

- Measures for controlling radiation risks must ensure that no individual bears an unacceptable risk of harm.

- People and the environment, present and future, must be protected against radiation risks.

- All practical efforts must be made to prevent and mitigate nuclear or radiation accidents.

- Arrangements must be made for emergency preparedness and response for nuclear or radiation incidents.

- Protective actions to reduce existing or unregulated radiation risks must be justified and optimized.

Nuclear Criticality Safety

Nuclear Criticality Safety was established as a discipline more than 50 years ago, in response to several accidents that had occurred in nuclear weapons programmes. The number of documented criticality accidents in "western" facilities over this period is slightly under 50. About 20 per cent of these accidents occurred in production plants, 10 per cent in working reactors (Chernobyl being the best known and the most dramatic accident) and the remainder in critical facilities where the properties of the assemblies themselves were being investigated. The rate at which these accidents occur has strongly decreased over the years. Such events have now become very rare.

The importance of the safe handling of all fissile materials was recognised at an early stage both by the scientific community and the responsible authorities. At the beginning, intensive experimentation with a large variety of configurations and materials took place in order to establish a basis of knowledge of such systems. At the time, computational methods and basic nuclear data were either not yet properly developed or had not reached sufficient sophistication to reliably predict the critical status of fissile materials.

Over the years, substantial experience has been gained in both experimentation and in data and code development. This state-of-the-art knowledge in criticality safety also has an economic impact. The reduction of uncertainties in safety margins allows improved and more economical designs for manipulation, storage and transportation of fissile materials.

New fuel cycles, handling of excess fissile materials from the weapons programmes and its possible use for civil energy applications make new demands on method development, experimentation and regulations.

This topic briefly describes the work carried out by the NEA in this field and outlines perspectives for the future.

The following activities have been carried out by the NEA on reactor fuel cycle safety:

- A publication entitled The Safety of the Nuclear Fuel Cycle gave a general overview of the safety issues and procedures in Member countries. It also included a chapter on regulatory issues.

- An incident reporting system specific to significant events from the fuel cycle has recently been set up: FINAS (Fuel Incident Notification and Analysis System).

- An international working group carries out a work programme on criticality safety.

Criticality Safety

Work has concentrated on the following studies of interest to Member countries:

- Criticality of nuclear fuel packages for transportation and storage;

- Criticality of fuel undergoing dissolution for reprocessing and partitioning;

- Burn-up credit criticality;

- Experimental data relevant to criticality safety.

Criticality of Nuclear Fuel Packages

A first study examined the ability of various computational methods to accurately compute criticality for systems which have been measured as experimentally critical. A procedure which evolved from this work allows assessing whether a given computational method produces "valid" results. This procedure provides a basis for acceptance of computational results by regulatory authorities on an international basis.

Criticality of Fuel Undergoing Dissolution

Eighteen experimental configurations were studied. In addition three calculational benchmarks on criticality codes for dissolving fissile oxides in acids have been completed. A particularly difficult problem proved to be the treatment of fuel double heterogeneities (solid fissile material surrounded by fissile material in solution).

Burn-up Credit

Burn-up credit is a term that applies to the reduction in reactivity of burned nuclear fuel due to the change in composition during irradiation. Conventional reactor codes and data used for in-core physics calculations can be used to evaluate the criticality state of burned, light water reactor (LWR) fuel. However, these codes involve complicated models and have large computational and data requirements. In reactor applications, these detailed analyses are required for the efficient operation of specific reactors. In away-from-reactor applications such as the design of casks (flasks) for the transportation of spent nuclear fuel, the candidate fuel for use in the cask may come from any reactor and it is desirable that the design allow for the inclusion of as much of the existing and expected fuel inventories as safely possible. In other words, for reactor operations the objective is to use most effectively very specific fuel in a specific application. For away-from-reactor applications the objective is a general design for a wide variety of fuels.

Traditionally, established away-from-reactor codes have been used for applications such as the design of storage and transportation (S/T) casks. In this type of analysis, the fuel is usually assumed to be at its full initial enrichment to provide a large safety margin for criticality safety analyses. The incentives for pursuing burn-up credit over the current, fresh fuel approach are widely recognised: the approach can extend enrichment limitations for existing S/T containers, and may contribute to the development of higher capacity S/T systems that would result in fewer fuel shipments and therefore decreased risk to the public. There is also potential application to criticality safety in dissolvers for fuel reprocessing as well as for timely and efficient transport to and from reprocessing facilities.

However, before such an approach can be approved by licensing agencies, it would be necessary to demonstrate that the available criticality safety calculational tools are appropriate for application to burned fuel systems and that a reasonable safety margin can be established. To this end, a suite of burn-up credit criticality benchmarks have been established by the NEA. The benchmarks have been selected to allow a comparison of results among participants using a wide variety of calculational tools and nuclear data sets. The nature of the burn-up credit problem requires that the capability to calculate both spent fuel composition and reactivity be demonstrated. The benchmark problems were selected to investigate code performance over a variety of physics issues associated with burn-up credit as described in table.

The focus here is the comparison of the results submitted by each participant to assess the capability of commonly used code systems, not to quantify the physical phenomena investigated in the comparisons or to make recommendations for licensing action. Participants use a wide variety of codes and methods using both deterministic and stochastic (Monte Carlo) techniques. Nuclear data were taken from several sources the Evaluated Nuclear Data Files (ENDF/B), the Japan Evaluated Nuclear Data Libraries (JENDL) and the Joint Evaluated Files (JEF).

Table: Summary of recent burn-up credit, benchmark problems addressed.

Benchmark	Primary objective
Phase I	• Examine effects of 7 major actinides and 15 major fission products for an infinite array of pressurised water reactor (PWR) rods at different burn-ups and cooling times. • Compare computed nuclide concentrations for depletion in a simple PWR pin-cell model to actual measurements at different burn-ups.
Phase II	• Examine the effect of axially distributed burn-up in an array of PWR pins as a function of initial enrichment, burn-up and cooling time. • Repeat study in a 3-D geometry representative of a conceptual burn-up credit transportation container.
Phase III	• Investigate the effects of moderator void distribution in addition to burn-up profile, initial enrichment, burn-up and cooling time sensitivities for an array of boiling water reactor (BWR) pins. • Compare computed nuclide concentrations for depletion in a BWR pin-cell model.
Phase IV	• Investigate burn-up credit for mixed oxide (MOX) spent fuel.

Experimental Data Relevant to Criticality Safety

In support of evaluation and validation of methods for criticality safety, several databases have been established:

Spent Fuel Isotopic Composition Database

A database of light water reactor (LWR) spent fuel assay data that has been compiled and contains isotopic inventory data collected from 13 LWRs, including 7 pressurised water reactors (PWRs) and 6 boiling water reactors (BWRs) in Europe, the United States and Japan as well as axial burn-up profiles.

International Handbook of Evaluated Criticality Safety Benchmark Experiments

This handbook (ICSBEP) contains criticality safety benchmark specifications that have been derived from experiments that were performed at various nuclear critical facilities around the world. The benchmark specifications are intended for use by criticality safety engineers to validate calculational techniques used to establish minimum subcritical margins for operations with fissile material. At present the handbook comprises about 2 000 critical configurations and is available on CD-ROM.

Criticality Benchmark Experiments

There have been several activities that involve experiments applicable to burn-up credit:

- Exponential Experiments in the Tank Typed Critical Assembly (TCA) of JAERI;

- International CERES Experimental Program designed for the validation of cross-section data and inventory predictions for actinides and fission products (France, UK and USA);

- Fission Product Experiments by the Institut de Protection et de Sûreté Nucléaire (IPSN), France;

- A set of experiments proposed by the United States called Spent Fuel Safety Experiment (SFSX) to provide integral benchmarks for validating spent fuel reactivity;

- Experiments on waste matrix materials planned at the Los Alamos National Laboratory, USA and at the Institute of Physics and Power Engineering, Obninsk, Russian Federation.

Working Party on Nuclear Criticality Safety

During a recent experts meeting on Needs for Critical Experiments, it was concluded that, in view of the limited number of operational critical facilities and the international scope of the needs for criticality safety technology, the NEA should encourage the performance of new critical measurements on a multilateral, international basis with regard to the sharing of facilities, staff expertise and funding resources. Also, given the absence of some experimental capabilities in many countries and the near-unique capabilities in others, the NEA should, through its Member countries representatives, recommend to their sponsoring agencies that certain facilities with unique capabilities be made available for international measurements programmes. This policy would reduce the need for redundancy in capabilities and promote stable funding for maintaining staff and equipment. With the expanding number and scope of NEA activities on criticality and safety, a Working Party for Nuclear Criticality Safety was established to provide guidance and overall co-ordination of these activities. This working party deals with technical, away-from-reactor, criticality safety issues relevant to fabrication, transportation, storage and other operations in the fuel cycle of nuclear materials. Specific Task Forces have been set up:

- Burn-up Credit Criticality (BUC),

- International Criticality Safety Benchmark Experiments Project (ICSBEP),

- Sub-critical Experiments (SUB),

- Experimental Needs in Criticality (EXN),

- Basic Minimum Values of Criticality (BMVC).

Additional areas and items of activity comprise:

- Criticality accidents (analysis, alarms, dosimetry),

- Databases (needs, developments, monitoring),

- Training and accreditation (exchange of national experience and approaches),

- Technical basis for standards,

- Nuclear data (liaison with Working Party on International Measurement Activities and Evaluation Co-operation),

- Codes and methods (Monte Carlo, probabilistic assessments, deterministic methods),

- Decommissioning (criticality principles - techniques, monitoring),

- Waste repository criticality issues.

Perspectives

If nuclear energy is to play an important role in our economies in the future (it currently represents 25 per cent of electricity production in the OECD area), fissile materials must be handled safely over the whole fuel cycle. Several fuel cycle options exist, and their advantages and disadvantages are being hotly debated at the technical, economical, political and public level. One of the questions being asked is: should plutonium be considered as a liability or an energy source?

In the coming years there will likely be further clarification of potential nuclear fuel cycle strategies, each one with its specific needs in criticality safety. Although a wealth of information is available from more than 50 years of cumulative knowledge acquired, case-specific analyses will be needed and will dominate criticality safety.

The release of fissile materials from some countries weapons programmes to the civil nuclear fuel cycle will influence future work according to available options. Open or closed fuel cycles, once-through or recycle will dominate the debate and research.

Diversity in national policies is expected to persist in this area in the short term. In the long term, and in order to leave the options open for future generations, it is worth sharing criticality information beyond national policies. It is for that reason that the Working Party on Nuclear Criticality Safety was set up, addressing issues that are of common interest to all countries with fuel cycle facilities. An international conference is scheduled for 1999 in France to address these themes as they relate to criticality safety, and to cover in particular the different fuel cycle options such as once-through, recycle and actinide partitioning, burning and transmutation.

Criticality safety calls for constant support and attention. A sound understanding and correct application of the principles of nuclear criticality safety are vital to the nuclear industry. The objective is to pursue an accident-free goal, while keeping in mind the repercussions that an avoidable criticality excursion could have. Current activities and future initiatives will obviously build upon past accomplishments. Events have shown that criticality safety is an international issue. It is therefore in the interest of all that information be widely shared and disseminated, notably through the NEA.

Nuclear Reactor Safety

Safety of a reactor is of prime concern to its owner for several reasons: to ensure the safety of the public, the reactor operators, and the investment itself. Therefore, the design of a reactor is developed according to industry standards.

These industry standards are developed by expert committees. They incorporate the best design, construction and operational standards, which have been developed over many decades of experience. For example, the standards for reactor pressure-vessel design arose out of standards for the construction of boilers for Mississippi River boats that had previously been failing. The standards have since been modified and added to for the fabrication of very thick stainless steel pressure vessels far advanced from those on the paddle-wheel steamers.

The building of a reactor takes many steps: choice of a suitable site; design of the power plant to fit that site; fabrication of the components and construction of the plant; low power commissioning; and, finally, full power operation. At each stage, the designer assures the safety of operation.

In addition, the licensing authority, the Nuclear Regulatory Commission (NRC), overviews the process and issues permits and licenses at significant points to allow work to proceed. The NRC has previously reviewed the industry standards and issued them (sometimes with modifications) as NRC standards incorporated into law — thereby ensuring that the best of industrial practices are incorporated into all new plants.

Design

Industry standards cover every aspect of plant design from its layout to the safety of individual components. For example, the construction of a 9" thick stainless steel vessel has its fabrication standards, while the wiring of monitoring and control circuits has its own standards. Each set of standards has been developed and approved by experts in that particular technology.

Most standards will incorporate certain safety principles, which are specified in general design criteria approved, again, by a consensus of experts.

Using these safety principles and the design standards for the details of design, fabrication and eventual construction, consequently results in a plant designed for safe operation.

Design Standards

A reactor is designed for a particular site according to industry standards for safe design and construction of all its components and systems as well as its operation. These industry standards incorporate the best design, construction and operational standards, which have been developed over many decades of experience. For example, the standards for reactor pressure-vessel design arose out of standards for the construction of boilers for Mississippi River boats that had previously been failing. The standards have since been modified and added to for the fabrication of very thick stainless steel pressure vessels far advanced from those on the paddle-wheel steamers. These standards are developed by expert committees sponsored by professional not-for-profit societies, such as the American Society of Mechanical Engineers.

Industry standards cover every aspect of plant design from its layout to the safety of individual components. For example, the construction of a 9" thick stainless steel vessel has its fabrication standards, while the wiring of monitoring and control circuits has its own standards. Each set of standards has been developed and approved by experts in that particular technology.

Many of the standards are very specialized, covering—for example—the details of dye-penetrant testing of pipes and vessels to search for any surface flaws during fabrication. Other standards deal with the separation of control systems from safety systems amongst the wiring, and still others deal with seismic response analysis of plant components. Most of the standards incorporate certain safety principles which are specified in the general design criteria approved by a consensus of experts.

Safety Principles

There are six principles of safety that are the same for all machinery, from modern cars to windmills to dams to nuclear plants. The goal in creating new machinery is to comply with as many of these principles as possible, depending on what the regulation for that particular machinery requires. In the nuclear industry, regulation requires compliance with all six of them.

These safety principles are:

- Multiple Barriers: Typically, if one knows that one might get hurt from machinery, the primary safety requirement is to keep the user clear of any danger by providing at least one or preferably multiple barriers. Cars offer multiple barriers in a similar fashion. They have collapsible front and rear ends, airbags, and seatbelts to protect people from harm in case of a crash. Nuclear power plants also offer multiple barriers: the canned fuel, the pressure vessel, and the containment to protect the public from release of radioactive materials.

- Redundancy: It is a good principle that if one of something must work, then more than one of them is included, just like there are four wheel brakes on a car. In a nuclear power plant, there are four different ways of obtaining a shutdown when needed: shutdown by heat feedback; dual active-shutdown systems; operator shutdown, or triple electronic systems.

- Diversity: Then, to be sure that one safety system works; it is preferable that the second system be of a different design and even made of different materials, just like foot brakes and hand brakes in a car. This ensures that if the first system fails, the second system won't fail for the identical reason. In a nuclear power plant, diversity is obtained through different designs of control rods and shutdown rods and through different designs of electronic systems.

- Protection from human error: While humans can act rapidly and responsibly, they can also make mistakes, so incorporating automatic actions for times when a human makes a mistake is good safety practice. For example, some cars offer automatic fuel line cut-off on impact. In nuclear power plants, an automatic reactor shutdown offers similar protection.

- Monitoring: One needs to know how the machinery is working and whether there are signs that it might fail, like having low pressure in your tires. Nuclear power plants are monitored extensively throughout the entire heat production process.

- Passive rather than active safety systems: Where possible, it is good practice to make use of the laws of nature to correct situations. In nuclear power plants, gravity is used to drop a barrier or a control rod more reliably than an active mechanism can do the same thing.

Designing a Reactor for Safe Operation

A reactor designer first makes sure that the fission process can be shutdown in a variety of ways. The primary defense is by ensuring that the core is designed so that high temperatures automatically change the core characteristics, fissioning is reduced, and temperatures go back down. This is called "inherent safety." Secondly, multiple automatic electronic systems monitoring the core for abnormal temperatures will insert control rods to stop the fission, and, if necessary, automatic fast shutdown rods are also available to act as brakes. Finally, there is always the operator who watches control room monitors. He can also instantly insert the shutdown rods from a single button.

Next, the designer must ensure that the reactor is always cooled, generally–in current commercial reactors–by water. Thus, the plant has more than one water path (usually three loops) to bring cooled water into the vessel. Monitors alert automatic valves to open or close so that adequate water is brought in and that the pressures in various vessels in the steam side are within limits. In an emergency, there is also a completely separate fourth cooling system. Finally, the operator also watches significant water temperatures, pressures and flows on monitors, to be able to take manual action if needed. This might include shutting the system down and activating the separate emergency cooling system. The operator can also bypass any valves that might be sticking.

The plant itself is designed to quality standards that are higher than in any other industry with sufficient margins in materials and designs to take care of any accident conditions. Materials do not melt just because the temperature is a little raised and vessels do not burst even when the pressure is above normal operational values.

Furthermore, the whole plant is designed to withstand earthquakes beyond any historically expected and to withstand extreme weather conditions and their consequences. For example, the outer containment can withstand hurricane missiles, such as a flying car or shafts of wood and metal. Swiss Green Party activists provided an excellent test of the containment when they fired mortar missiles at the French Superphénix plant from across the Rhône River. The missles only chipped the surface of the containment concrete. Later, during hurricane Andrew in Florida, the safest place for families of the operators as the hurricane passed directly over the plant was inside its containment.

Fabrication and Construction

As the design proceeds, the safety of the plant is analyzed thoroughly to meet and improve regulatory protection standards. The analysis is presented to the Nuclear Regulatory Commission as a Preliminary Safety Analysis Report (PSAR), and once that has been approved, the design is fixed. Then, the owner can proceed to fabrication and construction.

Since it takes many months to construct a nuclear plant, approval to fabricate and construct may be given in several steps depending on whether the item under consideration is safety-related or not. If an item is safety-related and appears as part of the PSAR submission, then it is kept under close regulatory scrutiny during fabrication.

Fabrication will take place at a number of locations: the vessels being built by one firm and electronics by another, and pumps and valves by others. However, all fabrication is performed to regulatory-approved industry quality standards. Material integrity of the large components, for example, is tested in several ways including radiographing and dye-penetrant methods; each governed by a standard. In this way, the owner can be assured that every piece of the plant is of high quality. The chain of construction will have no weak links.

Construction by the project's architect engineering firm is also performed to industry standards for such things as concrete mixes, rebar strength and density, welding techniques and so on. Construction is under close regulatory scrutiny, and there will be resident Nuclear Regulatory Commission inspectors on site while the work goes on.

Low-power Commissioning

After the plant design has been approved and the plant built, it must be tested before being placed into operation. This is called low-power commissioning, and it is a time to make corrections if anything proves to be out of specifications.

The Nuclear Regulatory Commission will permit vessels and piping to be filled with liquids (generally water), pumps can be operated and valves can be opened and closed. This is especially so for emergency systems that, although never expected to be operated in the plant's lifetime, must operate if called upon. The process is similar to the first fill of a newly-installed replacement car engine with oil and the first cranking of the pistons. Although you are not going to drive it a hundred-miles-and-hour today, you would like to know that it all works and nothing leaks.

Naturally, the inspectors of the Nuclear Regulatory Commission are involved throughout these low-power commissioning tests. When the plant operators and the regulators are satisfied with all the tests and that they agree with predictions (which go into the safety case), the plant may be granted a license for full-power operation.

Full-power Operation

Full-power operation is a natural extension of low-power commissioning. It is exactly like starting and running your car when you buy it, since you know that all the low-power commissioning tests have been done.

Sometimes, the Nuclear Regulatory Commission may license the plant to come to power in a series of steps: 25%, 50%, 90% and 100% of full power, but the effect is the same. The result is full-power operation and the quiet generation of clean electricity for the plant lifetime of 30 to 40 years or more.

Other Items Contributing to Safety

Safety is a comprehensive state encompassing many things including good management, safe design, industry standards, and positive regulation. Safety also includes well-trained staff and operators, attention to emergency plans if anything went wrong, and, on another level, security.

Training

Training is vital to ensure that operators and other staff such as maintenance crews know exactly what to do and why they are doing it that way. In addition, training ensures that everyone is trained to do the job the same way.

Therefore, staff training at all levels is on-going. Operators must go through yearly training (with exams), and success in training is necessary for them to continue in their position. It would not be too strong to say that good training of all staff, at all levels, is a basis for safe operation.

Emergency Plans

While it is very unlikely that anything will go seriously wrong, emergency plans are set in place to protect operators and anyone who might be in the plant or the vicinity. These emergency plans involve close cooperation with off-site agencies like the police, the fire department, and even school buses, if they are part of any evacuation plan that has been approved.

Security

Security is related to safety including and beyond plant operation. Therefore, security plans are confidential and known only to those with a need to know: plant management, the police, certain regulators, and anyone with a role in response.

One part of security is aimed at individuals who might attempt to disrupt the operation of the plant or steal materials from it. Neither has ever been done, nevertheless security plans are made for each plant. They do not contribute directly to safety of operation but they would stop anyone, like a terrorist, who might have harmful objectives.

Nuclear Power Plant Safety Systems

Each nuclear power plant in Canada has multiple, robust safety systems designed to prevent accidents, and reduce its effects should one occur. All of these systems are maintained and inspected regularly, and upgraded when necessary, to ensure plants meet or exceed strict safety standards established by the Canadian Nuclear Safety Commission. The systems perform three fundamental safety functions: controlling the reactor, cooling the fuel and containing radiation.

How a Nuclear Power Plant Works

Reactor

All nuclear power plants in Canada use the CANDU design - a safe, reliable, reactor technology.

CANDU reactors produce electricity through a process known as fission. Fission is the process of splitting atoms of natural uranium inside the reactor, releasing radiation and heat.

The split atoms then continue a "chain reaction": more atoms continue to be split, resulting in more radiation and heat.

The heat - energy - is harnessed to make steam to power the turbines and generators, which in turn produce electricity.

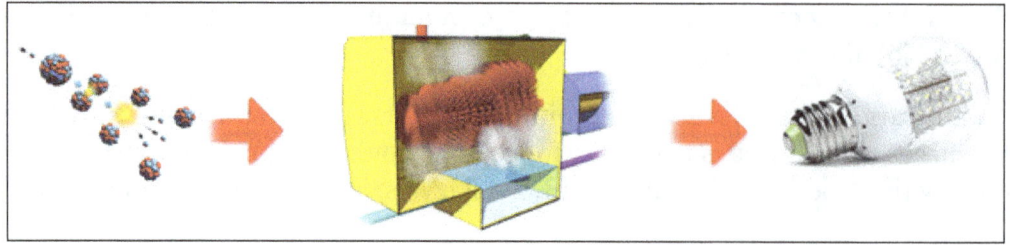

Used Nuclear Fuel Pool

After the uranium, or nuclear fuel, has been used in the reactor, it is removed and stored securely in a pool for a period of 6 to 10 years. The water in the pool continues to cool the fuel and provides shielding against radiation.

All of Canada's fuel pools are built in ground, in separate buildings at the nuclear power plant, and are designed to withstand earthquakes.

Fuel pool.

Controling the Reactor

Normal Operation

Controling the reactor involves increasing, decreasing or stopping the chain reaction happening inside the reactor.

When the reactor is operating, the chain reaction (or power level) is controlled by moving adjuster rods and varying the water level in vertical cylinders.

Sensitive detectors constantly monitor different aspects, like temperature, pressure and the reactor power level.

When necessary, CANDU reactors can safely and automatically shut down within seconds.

Shutdown Systems

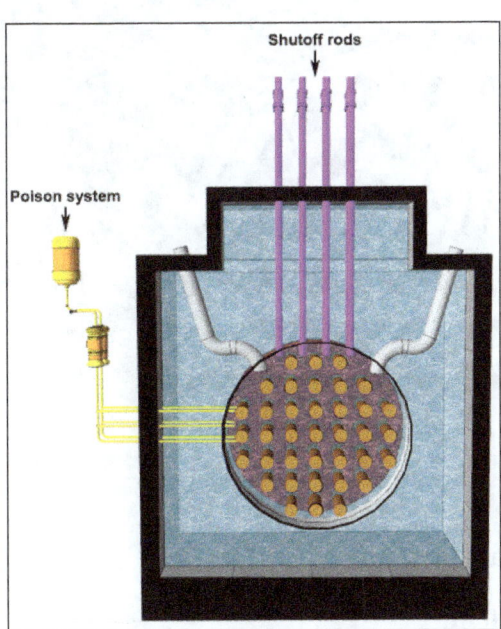

All nuclear power reactors in Canada have two independent, fast-acting and equally effective shutdown systems.

The first shutdown system is made up of rods that drop automatically and stop the chain reaction if something irregular is detected.

The second system injects a liquid, or poison, inside the reactor to immediately stop the chain reaction.

Both systems work without power or operator intervention. However, they can also be manually activated. These systems are regularly and safely tested.

Restarting the Reactor

Once a CANDU reactor is shut down, it will stay that way until restarted by the operators in the control room.

There is no possibility of the reactor accidentally restarting on its own after it's shut down. The reactor must be manually restarted. This is another important safety feature.

Cooling the Fuel

Decay Heat

Following shutdown, the amount of energy produced by the reactor decreases rapidly.

The nuclear fuel will, however, continue to produce some heat and must be cooled.

That heat, called decay heat, represents a small fraction of the heat produced during normal operation.

CANDU fuel bundle.

Main Cooling Systems

Fuel cooling involves three main systems:

- The heat transport system,

- The steam system,

- The condenser cooling system.

The heat transport system brings the heat produced by the reactor to the steam generators.

This system is made up of very robust pipes, filled with heavy water - a rare type of water found in nature. Pipes and other components are maintained and inspected regularly, and replaced if necessary.

Inspections include measuring pipe wear and tear and identifying any microscopic cracks or changes well before they become a problem.

On average, one out of 7,000 drops of water is heavy water. It is 10% heavier than regular water because it includes a heavy form of hydrogen called deuterium.

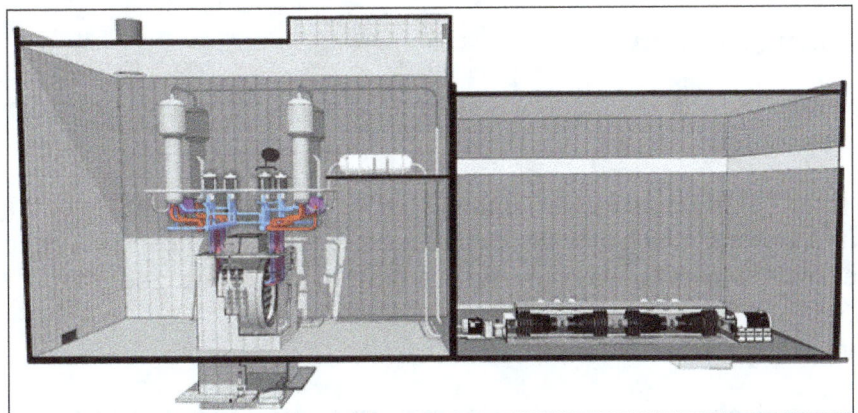

Heat transport system.

The second system, the steam system, uses normal water. The heat from the reactor turns this water into steam to run the turbines and generators.

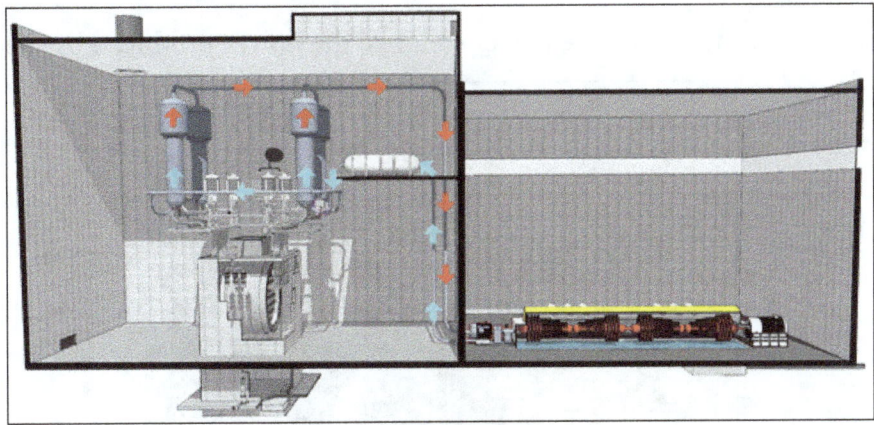

Steam system.

That steam is then cooled and condensed using a third system that pumps in cold water from a body of water such as a lake or reservoir. This is called the condenser cooling system.

Like other components, the steam and condenser cooling systems are regularly inspected.

These inspections take place throughout the life of the nuclear facilities to confirm that aging equipment is functioning as originally designed.

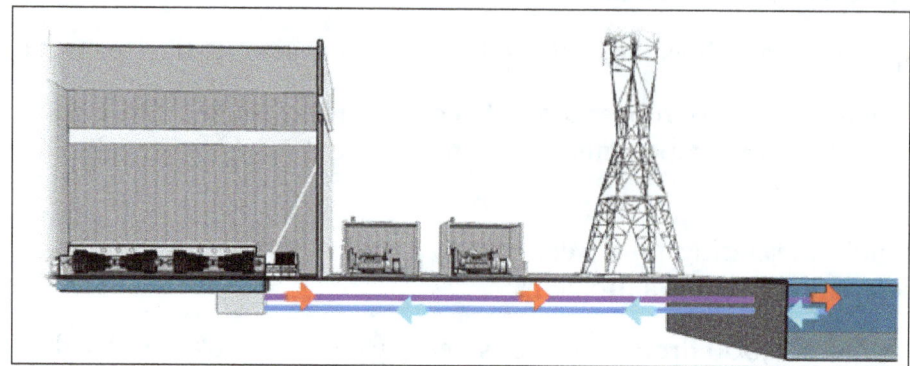

Condenser cooling system.

Shutdown Cooling System

A simpler cooling system is used when the reactor is shut down for an extended period, for example during a planned outage.

It requires little power to function and is connected directly to the heat transport system. It allows the primary coolant system to be partly drained to perform inspection and maintenance work (e.g., inspection of the steam generator tubes or replacement of pump components).

Multiple Power Supplies

Cooling systems need electricity to operate. Under normal operation, they get their electricity from the same power grid as the rest of us.

Nuclear power plants in Canada are also equipped with multiple sources of backup power if they get disconnected from the grid.

Sources of backup power include onsite power - that is, the power produced by the plant itself.

Emergency power generators.

In addition, the following are available:

- Two or three standby power generators,

- Two or three emergency power generators,

- Emergency batteries.

Some plants include even more equipment.

You can learn more by watching what would happen in the very unlikely event of a total station blackout – the situation that led to the Fukushima accident following the large tsunami that destroyed all available power sources onsite.

Natural Circulation

One of the inherent and proven safety features of CANDU reactors is their ability to cool the reactor through natural circulation.

In CANDU reactors, natural circulation takes over when the pumps that normally push the coolant through the heat transport system stop functioning.

For natural circulation to continue over time, steam generators need to be filled with cool water.

How does it Work?

This cooling feature of CANDU reactors works because of the difference in temperature and elevation between the steam generators (cooler and physically higher than the reactor core) and the reactor core (hotter and lower than the steam generators).

Emergency Injection Systems

Emergency pressurized nitrogen tanks.

In the unlikely event of a loss of heavy water, which could, for example, be caused by a pipe break, the emergency injection system would ensure water continues to circulate over the containers holding the fuel to cool it.

They would do this by working with pressurized tanks of nitrogen or pumps.

A collection basin located in the basement of the reactor building would recover the water and pump it back into the reactor until repairs could be made.

Emergency Mitigation Equipment

CNSC inspector verifying a portable emergency power generator.

As one of the actions mandated by the CNSC following the Fukushima accident, nuclear power plant operators in Canada have been acquiring emergency mitigation equipment, such as portable power generators and pumps, which could be used to bring reactors to a safe shutdown state during a severe accident.

The equipment, located onsite and offsite, is easily transported and could be used in several ways.

For instance, it could be used to stabilize reactors, supply power to the control room and add water to the used nuclear fuel pools so they could continue cooling the used nuclear fuel.

Containing Radiation

Containment Layers

Nuclear reactors are built with multiple barriers to safely contain radiation. At the heart of all CANDU reactors are hardened ceramic pellets made of natural uranium.

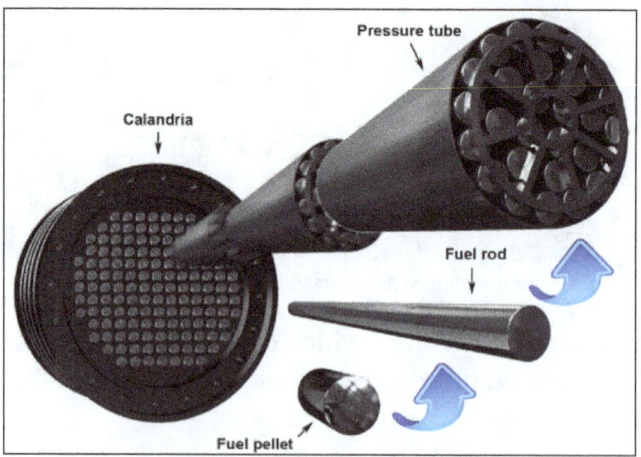

These pellets contain the radiation. They form the first layer of containment. The pellets are enclosed in rods, which form the second layer of containment. CANDU fuel rods are made of zircaloy, a metal alloy extremely resistant to heat and corrosion.

The rods are then loaded into pressure tubes, which are part of the heat transport system. This is the third layer of containment. The pressure tubes are contained inside a metal tank called the calandria, which itself is contained inside a thick vault made of reinforced concrete.

The fourth layer of containment is the building that houses and protects the reactor.

The walls of the reactor building are made of at least one metre of reinforced concrete.

The reactor building is surrounded by an exclusion (buffer) zone.

Minimizing Radiation Releases

During normal operation, nuclear power plants release very small amounts of radiation into the air and water.

These releases come from the reactor and its system and from waste management activities.

In order to reduce airborne releases, highly efficient filters and radiation monitors are installed as part of the ventilation systems.

Filters remove over 99% of the radiation from the air before it is released to the environment. Similar systems are also installed to remove radioactivity from waterborne releases.

CNSC inspector verifying radiation levels.

These releases usually come from wash water used to clean surfaces, floors and laundry, as well as from water draining from showers and sinks.

All radiation releases from nuclear facilities in Canada are very small. They are monitored and controlled by the plant operator, and reported to the CNSC.

The release levels are well below regulatory limits and do not pose any risk to the health and safety of persons or the environment.

Filtering systems are regularly inspected and power plant operators must, by law, report all radioactive releases into the environment.

Protecting Containment in Case of an Accident

Safety systems are in place so that, in case of an accident, they can protect the containment from internal pressure due to steam releases inside the reactor building.

In a single-unit station, internal pressure would be lowered by spraying water from a dousing tank.

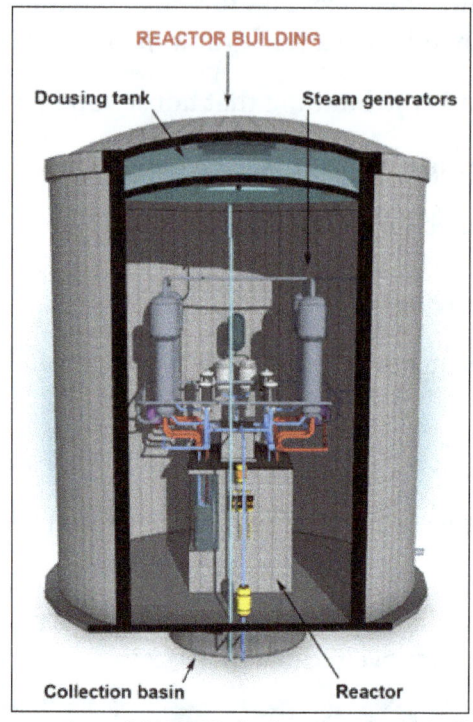

Cutaway view of CANDU single unit reactor building.

In a multi-unit station, pressure would be lowered by releasing steam and hot gases from the reactor building to the vacuum building.

The vacuum building is a structure specifically designed to quickly and safely lower pressure inside the reactor building. This building also has a dousing system to control pressure.

The vacuum and dousing systems work without power and are tested periodically under the supervision of CNSC inspectors.

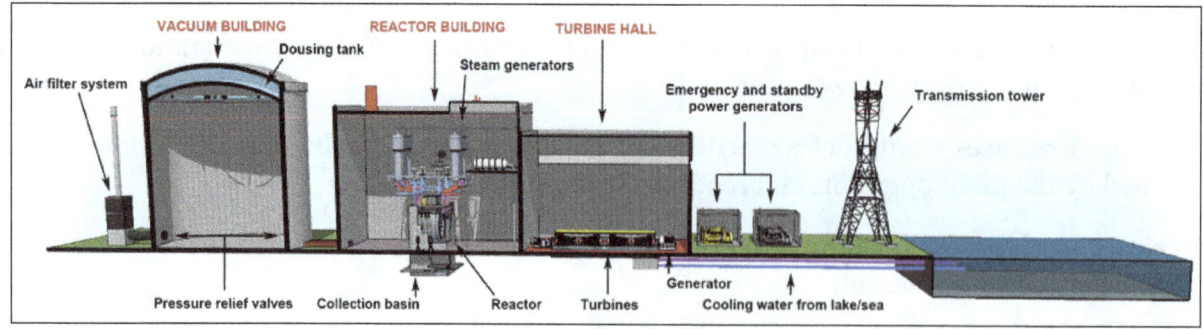

Cutaway view of a CANDU multi-unit nuclear power plant.

Hydrogen Management

Hydrogen gas can be produced during a severe accident. Hydrogen gas, which is flammable, could cause an explosion and damage the containment, as well as to personnel and other parts of the plant.

CNSC inspector taking a first-hand look at a newly
installed passive autocatalytic recombiner.

To deal with the potential hazards of hydrogen gas, most CANDU plants are equipped with hydrogen igniters or burners.

Recently, nuclear power plant operators have begun installing passive autocatalytic recombiners.

These are devices which passively (without need for external power) remove hydrogen from the containment and effectively reduce the risk of an explosion or fire.

References

- Disasters-nuclear-accidents, educational-magazines: encyclopedia.com, Retrieved 16 June, 2019

- What-is-nuclear-safety: suredyna.com, Retrieved 26 January, 2019

- Reactor-safety, know-nuclear-talking-nuclear: nuclearconnect.org, Retrieved 15 March, 2019

- Nuclear-power-plant-safety-systems, power-plants: nuclearsafety.gc.ca, Retrieved 19 February, 2019

- M.C. Brady et al.: "International Studies on Burnup Credit Criticality Safety by an OECD/NEA Working Group, Proc. International Conference on the Physics of Nuclear Science & Technology, Long Island, NY, USA, 5-8 October 1998

Permissions

All chapters in this book are published with permission under the Creative Commons Attribution Share Alike License or equivalent. Every chapter published in this book has been scrutinized by our experts. Their significance has been extensively debated. The topics covered herein carry significant information for a comprehensive understanding. They may even be implemented as practical applications or may be referred to as a beginning point for further studies.

We would like to thank the editorial team for lending their expertise to make the book truly unique. They have played a crucial role in the development of this book. Without their invaluable contributions this book wouldn't have been possible. They have made vital efforts to compile up to date information on the varied aspects of this subject to make this book a valuable addition to the collection of many professionals and students.

This book was conceptualized with the vision of imparting up-to-date and integrated information in this field. To ensure the same, a matchless editorial board was set up. Every individual on the board went through rigorous rounds of assessment to prove their worth. After which they invested a large part of their time researching and compiling the most relevant data for our readers.

The editorial board has been involved in producing this book since its inception. They have spent rigorous hours researching and exploring the diverse topics which have resulted in the successful publishing of this book. They have passed on their knowledge of decades through this book. To expedite this challenging task, the publisher supported the team at every step. A small team of assistant editors was also appointed to further simplify the editing procedure and attain best results for the readers.

Apart from the editorial board, the designing team has also invested a significant amount of their time in understanding the subject and creating the most relevant covers. They scrutinized every image to scout for the most suitable representation of the subject and create an appropriate cover for the book.

The publishing team has been an ardent support to the editorial, designing and production team. Their endless efforts to recruit the best for this project, has resulted in the accomplishment of this book. They are a veteran in the field of academics and their pool of knowledge is as vast as their experience in printing. Their expertise and guidance has proved useful at every step. Their uncompromising quality standards have made this book an exceptional effort. Their encouragement from time to time has been an inspiration for everyone.

The publisher and the editorial board hope that this book will prove to be a valuable piece of knowledge for students, practitioners and scholars across the globe.

Index

www.ingramcontent.com/pod-product-compliance
Lightning Source LLC
Chambersburg PA
CBHW080404190526

45161CB00003B/123